PRACTICAL
PROBLEMS in
MATHEMATICS
for ELECTRICIANS

Sixth Edition

PRACTICAL PROBLEMS in MATHEMATICS for ELECTRICIANS

Sixth Edition

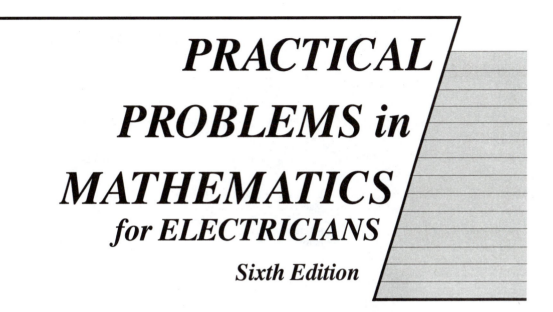

Stephen L. Herman
and
Crawford G. Garrard

DELMAR

THOMSON LEARNING ™

Australia   Canada   Mexico   Singapore   Spain   United Kingdom   United States

DELMAR

THOMSON LEARNING

Practical Problems in Mathematics for Electricians, 6ᵗʰ edition

By
Stephen L. Herman
and
Crawford G. Garrard

Business Unit Director:
Alar Elken

Executive Editor:
Sandy Clark

Senior Acquisitions Editor:
Gregory L. Clayton

Development Editor:
Jennifer A. Thompson

Executive Marketing Manager:
Maura Theriault

Marketing Coordinator:
Karen Smith

Executive Production Manager:
Mary Ellen Black

Production Manager:
Larry Main

Production Coordinator:
Andrew Crouth

Project Editor:
Christopher Chien

Art/Design Coordinator:
David Arsenault

Cover Illustration:
John Kenific

Library of Congress Cataloging-in-Publication Data:

Herman, Stephen L.
   Practical problems in mathematics for electricians / Stephen L. Herman and Crawford G. Garrard.—6th ed.
      p. cm.—(Delmar's practical problems in mathematics series)
   ISBN 0-7668-3897-8 (alk. paper)
   1. Electric engineering—Mathematics. I. Garrard, Crawford G. II. Title. III. Series.

TK153 .H44 2001
513'.13'0246213—dc21

2001042465

NOTICE TO THE READER

Contents

Preface

Practical Problems in Mathematics for Electricians, Sixth Edition, is one in a series of widely used books designed to offer students, trainees, and others with practical problem-solving experience in a particular trade area. The sixth edition contains added explanations to some units to aid in understanding how certain mathematical operations are performed. This edition also contains a new unit on scientific notation, and explains engineering units used in most scientific and engineering applications. Most of the problems are word problems and many contain multiple steps. These problems are intended to exercise the reasoning ability of students.

Practical Problems in Mathematics for Electricians offers many benefits for both instructors and students. The book begins with addition and subtraction of whole numbers and progresses through basic algebra and trigonometry. It is an excellent supplement to any mathematics textbook in which more detailed treatment of appropriate mathematics topics can be found. Students of electricity will find excellent opportunities to test and develop their problem-solving abilities, and will receive a valuable review of electrical terminology in the process.

A glossary is included to aid students with technical and mathematical definitions, and the appendix provides information on measurement, formulas, and trigonometric functions. Answers to odd-numbered problems are also provided.

Instructor's Guide to Accompany Practical Problems in Mathematics for Electricians provides the answers to all problems and contains two achievement reviews to provide an effective means of measuring student progress.

ABOUT THE AUTHORS

Stephen L. Herman is currently lead instructor in the Industrial Electricity Department at Lee College in Baytown, Texas. He has extensive experience as a teacher and an industrial electrician and is the author of textbooks for students of the electrical trades.

The late Crawford G. Garrard taught at the Augusta Area Technical School in Georgia and was active in the field of technical education.

REVIEWERS

The authors and publisher wish to thank the following individuals whose thorough review of the revision manuscript helped to create a textbook that is both technically accurate and complete:

David Alimena
Exxon Mobil Inc.
Baytown, TX

Dewain Belote
Hillsbrough Community College
Tampa, FL

Jay Blumenthal
Kansas City Community College
Kansas City, KS

Wes Mozley
Wesley Enterprises
Albequerque, NM

To the Student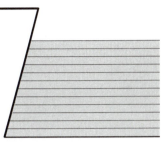

SAFETY PRACTICES FOR ELECTRICIANS

1. Be sure the power in the circuit is disconnected. An electrical circuit is safe only after it has been unplugged and the capacitors have been discharged.

2. Be sure your hands are dry when handling an electrical or electronics device that is connected to a live power line. Perspiration may cause your hands to be damp. Any amount of moisture on the surface of the skin reduces resistance to current, thereby increasing the danger of severe shock.

3. Use only those tools and equipment that are in good condition. Replace defective cords and plugs immediately.

4. Use only those electrical or electronic devices that have been approved by the Underwriters Laboratories, Incorporated. This will ensure the maximum degree of safety under the conditions for which the product was designed to operate.

5. Use only carbon dioxide or dry chemical fire extinguishers for the control of fires involving electrical equipment that is connected to power lines.

6. Work on an energized circuit with only one hand. Keep the other hand behind your back or in a pocket. This procedure will prevent current from passing through the chest region of your body.

7. Use an isolation transformer in conjunction with the operation of AC–DC equipment. This procedure reduces the danger of shock and damage to instruments.

8. Be sure to have another person nearby when working on equipment that may be electrically "hot."

9. Be sure to stand on a dry surface when using electrical or electronic devices. Rubber-soled shoes or a plastic mat will block the path of electricity to the ground.

10. Use caution and common sense when working with electricity.

Whole Numbers

 ## Unit 1   ADDITION OF WHOLE NUMBERS

BASIC PRINCIPLES OF ADDITION OF WHOLE NUMBERS

Whole numbers are numerical units with no fractional parts. Addition is the process of finding the *sum* of two or more numbers. Whole numbers are added by placing them in a column with the numbers aligned on the right side of the column. The right column of numbers is added first. The last digit of the sum is written in the answer. The remaining digit is carried to the next column and added. This procedure is followed until all columns have been added.

Example: Find this sum. 25 + 7 + 126 + 54 + 367

2	*12*	*12*
25	25	25
7	7	7
126	126	126
54	54	54
+ 367	+ 367	+ 367
9	79	579

PRACTICAL PROBLEMS

1. When taking inventory, you find that the numbers of BX connectors in five different bins are 176, 264, 375, 234, and 116. What is the total number of connectors in all bins?

2. Eight different boxes contain a number of ¾-inch, #8 flat-head, bright wood screws. The numbers of screws are 124, 72, 36, 92, 38, 64, 74, and 67. What is the total number of screws?

3. In wiring eight houses, you are to install outlets. The graph shows the number
    of outlets to be installed in each house. Find the total number of outlets that
    must be roughed-in.

    _____

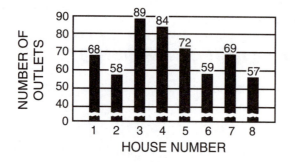

4. An electrician uses switch outlet boxes on eight different jobs. The number of
    boxes used on each job is 56, 9, 86, 36, 93, 105, 42, and 56. Find the total
    number of outlet boxes used.

    _____

5. The materials charged to a wiring job are as follows: 100-ampere distribution
    panel, $118; meter switch, $38; conduit, $64; number 2 wire, $88; BX cable,
    $73; conduit fittings, $26; outlet boxes, $153; switches, $112; fixtures, $215;
    and $64 for wire nuts, grounding clips, staples, and pipe clamps. What is the
    total amount charged for these materials?

    _____

6. At different times during a week, an electrician takes the following amounts
    of metallic cable from stock: 500 feet, 1,200 feet, 250 feet, 90 feet, 38 feet,
    65 feet, 84 feet, 225 feet, and 125 feet. What is the total number of feet of
    metallic cable taken from stock?

    _____

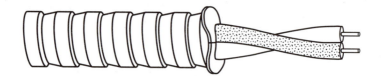

7. The following amounts of nonmetallic cable are used on an apartment house
    job: 625 feet, 785 feet, 75 feet, 140 feet, 310 feet, 325 feet, and 120 feet.
    What is the total number of feet of nonmetallic cable used on the job?

    _____

8. A factory department has motors of 75 horsepower, 30 horsepower, 200 horsepower, 40 horsepower, 25 horsepower, 15 horsepower, 5 horsepower, 125 horsepower, 150 horsepower, and 175 horsepower. What is the combined horsepower of the 10 motors? _____

9. An electrical supply house purchases solder in separate lots of 35 pounds, 40 pounds, 125 pounds, 200 pounds, 75 pounds, 90 pounds, 20 pounds, and 30 pounds. Find the total number of pounds of solder purchases. _____

10. The line graph shows the monthly consumption of energy in kilowatt-hours for a house during a one-year period. Find the total amount of energy consumed during the year. _____

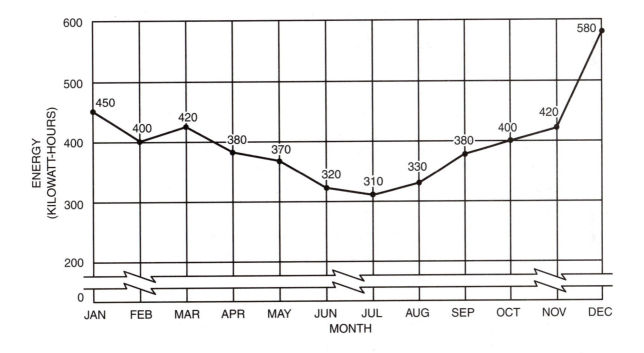

11. A school has twelve lighting circuits that use the following wattages: 545 watts, 650 watts, 750 watts, 1,820 watts, 2,462 watts, 2,571 watts, 1,360 watts, 1,540 watts, 793 watts, 1,225 watts, 330 watts, and 793 watts. What is the total number of watts consumed when all these circuits are being used? _____

12. The cost of magnet wire for a motor repair shop during a one-week period is as follows: 14 pounds of number 17, $58; 12 pounds of number 16, $55; 10 pounds of number 24, $51; 6 pounds of number 21, $19; 5 pounds of number 25, $24. Find the total cost of magnet wire during this period. _____

13. From a full container of dry cells, 325 dry cells are placed in the stockroom, 45 dry cells are placed on the shelf in the showroom, and 18, 25, 30, 24, and 6 dry cells are sold to customers. How many dry cells are taken from the full container? _____

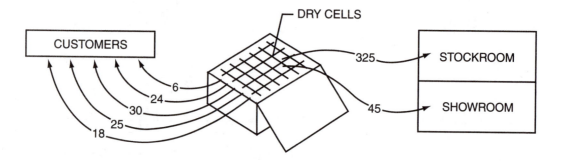

14. Three rooms of a house contain lamps that have the following wattage: living room, 150 watts; dining room, 125 watts; bathroom, 75 watts. What is the total load when all lamps are operating? _____

15. The following number of BX cable staples are used during a given period: 250, 125, 65, 36, 48, 96, 92, 28, 42, 106, 140, and 24. Find the total number of BX cable staples used during this period. _____

16. An electrical contractor receives the following quantities of Braidx cable during the first quarter of the year: January, 7,500 feet; February, 10,750 feet; March, 4,500 feet. Find the total number of feet of Braidx cable received. _____

17. During one week of work, an electrician uses the following amounts of 3-conductor wire with ground NM cable: 1,200 feet, 1,150 feet, 1,076 feet, 180 feet, and 100 feet. Find the total amount of cable used. _____

 # Unit 2 SUBTRACTION OF WHOLE NUMBERS

BASIC PRINCIPLES OF SUBTRACTION OF WHOLE NUMBERS

Subtraction is the process of finding the *difference* between two numbers. The smaller of the two numbers is placed below the larger, keeping the right column of numbers aligned.

Example: Subtract 432 from 768.

$$
\begin{array}{r}
768 \\
-\ 432 \\
\hline
336
\end{array}
$$

BORROWING

In subtracting whole numbers, it is sometimes necessary to borrow from the number in the adjacent column. When you do this, the amount borrowed must be in increments of value of the column borrowed from. For example, starting from the right, the first column represents units or 1s, the second column represents 10s, the third column represents 100s, the fourth column represents 1,000s, and so on. The number 9,876 could actually be rewritten as 1,000 nine times, 100 eight times, 10 seven times, and 1 six times.

1,000	100	10	1
1,000	100	10	1
1,000	100	10	1
1,000	100	10	1
1,000	100	10	1
1,000	100	10	1
1,000	100	10	
1,000	100		
1,000			
9,000	800	70	6

Now assume that the number 7,787 is to be subtracted from 9,876.

Write the smaller number below the larger.

$$
\begin{array}{r}
9,876 \\
-7,787 \\
\hline
\end{array}
$$

In this example, 7 cannot be subtracted from 6. Therefore, the 6 must borrow from the 7 in the column adjacent to it. Since the 7 is in the 10s column, 10 is borrowed, leaving 60 in that column. The borrowed 10 is added to the original 6, making 16 (10 + 6 = 16). Now 7 can be subtracted from 16, leaving a difference of 9.

$$
\begin{array}{r}
986 \ \textbf{(16)} \\
-778 \ (\ \textbf{7}) \\
\hline
9
\end{array}
$$

In the next column, 80 must be subtracted from 60. Since this is not possible, 100 will be borrowed from the 800 in the adjacent column and added to the existing 60, making 160. (This now leaves 7 in the 100s column.) The difference will be 80.

$$
\begin{array}{r}
97 \ \textbf{(160)} \ 6 \\
-77 \ (\ \textbf{80}) \ 7 \\
\hline
8 \quad 9
\end{array}
$$

In the third column, 700 is subtracted from 700, leaving a difference of 0.

$$
\begin{array}{r}
9{,}876 \\
-7{,}787 \\
\hline
089
\end{array}
$$

In the fourth column, 7,000 is subtracted from 9,000, leaving a difference of 2,000.

$$
\begin{array}{r}
9{,}876 \\
-7{,}787 \\
\hline
2{,}089
\end{array}
$$

THE "BORROW 1" METHOD

"Borrow 1" is another borrowing method used in the subtraction of whole numbers. The term is actually a misnomer because 1 can be borrowed only from the units column, but many people use this method and find it simpler to understand. Assume the number 58 is to be subtracted from the number 843. Place 58 below 843.

$$
\begin{array}{r}
843 \\
-\ 58 \\
\hline
\end{array}
$$

Since 8 cannot be subtracted from 3, 1 is borrowed from the adjacent 4. The 3 now becomes 13, and the 4 becomes 3.

$$\begin{array}{r} 83\ (13) \\ -\ \ 5\ 8 \\ \hline 5 \end{array}$$

The 5 must now be subtracted from the 3 in the second column. Since 5 cannot be subtracted from 3, 1 is borrowed from the adjacent 8, and the 3 becomes 13. The 8 now becomes a 7.

$$\begin{array}{r} 7\ (13)\ \ 3 \\ -\ \ \ \ 5\ 8 \\ \hline 7\ \ 8\ 5 \end{array}$$

PRACTICAL PROBLEMS

1. An electrician removes from stock 500 feet of BX cable on Monday, 250 feet on Tuesday, and 750 feet on Wednesday. On Friday, 339 feet of BX cable are returned. How many feet of BX cable are used? _____

2. An electrical contractor charges $598 for a job. The materials cost $263. The cost of labor is $173, and the cost of transportation is $10. Find the profit. _____

3. An inventory sheet shows 565 outlet boxes on January 1. On the 10th of January, 145 boxes are taken out of stock. On the 14th of January, 35 boxes are returned to stock. How many outlet boxes are in stock after January 14? _____

4. For a residential job, a reel containing 1,050 feet of cable is delivered. Three 45-foot lengths and three 65-foot lengths are used. How many feet are left? _____

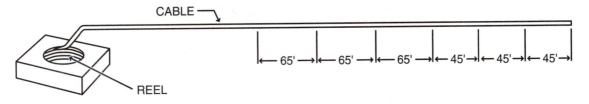

5. A coil of Type S cord, 250 feet long, is taken on a job. The lengths cut off for drop and extension cords are 30 feet, 15 feet, 8 feet, 25 feet, 15 feet, and 20 feet. How many feet of cord remain in the coil? _____

6. A 1,000-foot reel of large stranded cable weighs 1,106 pounds. Of this, 365 pounds are used on a certain job from the switch to the first pull box, and 422 pounds are used from the first box to the last box. How many pounds of wire are left on the reel?

_____

7. A purchase of 2,500 feet of number 14 wire is made for a job. On November 1, 1,365 feet of this wire are used. On November 3, an additional 830 feet are used. How many feet of wire are left after November 3?

_____

8. On a certain job, a sum of $438 is spent for materials. Of this amount, $76 is spent for 1-inch conduit, and $105 is spent for cable. How much money is spent for other materials?

_____

9. During the month of December, 400 outlet boxes are purchased at a cost of $385. The numbers of outlet boxes used are as follows: 59 boxes on December 1, 69 boxes on December 5, and 72 boxes on December 12. How many outlet boxes are left?

_____

10. At inventory time, 435 pounds of magnet wire for winding motors are checked as being available. In ten successive days, 15 pounds, 6 pounds, 24 pounds, 12 pounds, 3 pounds, 8 pounds, 17 pounds, 32 pounds, 16 pounds, and 13 pounds of wire are taken out of stock. How many pounds are left?

_____

11. A customer receives an electricity bill. The bill states that 1,876 kilowatt-hours of energy are used. Of this total, 504 kilowatt-hours are used for lighting and the rest are used for hot water. How many kilowatt-hours does the customer use for hot water?

_____

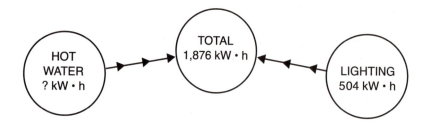

12. A supply house has 804 solenoids for 5-horsepower motor controls. The clerk must reorder this item when the supply reaches 60. Fourteen are sold on January 1, and 75 are sold on January 16. How many more can be sold before reordering?

_____

13.   A bin contains a total of 173 octagon boxes. For two jobs, 47 boxes and 65 boxes are taken from the bin. One job uses 4 boxes less than originally estimated, and these 4 are returned to the bin. How many boxes are in the bin at the end of the two jobs?   _____

14.   A buyer can purchase 70 screwdrivers. Ten 4-inch lengths, twelve 6-inch lengths, twenty 8-inch lengths, and twenty 10-inch lengths are needed. How many heavy 24-inch length screwdrivers can be bought to obtain the total of 70 screwdrivers?   _____

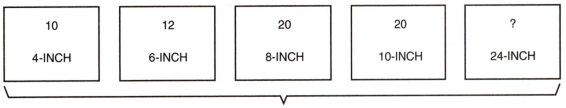

| 10 | 12 | 20 | 20 | ? |
| 4-INCH | 6-INCH | 8-INCH | 10-INCH | 24-INCH |

TOTAL = 70 SCREWDRIVERS

15.   An electrician is given a 250-foot coil of BX cable. On one run, 29 feet are used. How much BX cable is returned to stock?   _____

16.   A two-wire transmission line requires 134 miles of conductor for each wire. By straightening the proposed right-of-way, the distance is reduced by 7 miles. What will be the length of each new wire?   _____

17.   A total resistance of 60 megohms is needed. On hand are three resistors with the following values: 14 megohms, 25 megohms, and 11 megohms. What is the value of the additional resistor required?   _____

18.   A tapered pin has a small-end diameter of 101 centimeters and a large-end diameter of 189 centimeters. What is the difference between the two diameters?   _____

Unit 3   MULTIPLICATION OF WHOLE NUMBERS

BASIC PRINCIPLES OF MULTIPLICATION OF WHOLE NUMBERS

Multiplication is actually a method of addition used when like numbers are added. For example, if four 5s are added, the answer will be 20. If the number 5 is multiplied by 4, the answer (known as the *product*) is equal to 20. Therefore, 5×4 is the same as adding four 5s.

$$
\begin{array}{r}
5 \\
5 \\
5 \\
+5 \\
\hline
20
\end{array}
\qquad
\begin{array}{r}
5 \\
\times 4 \\
\hline
20
\end{array}
$$

To multiply larger numbers, first write the number to be multiplied; then write underneath it the number of times it is to be multiplied. In the following example, the number 247 is to be multiplied by 32. Write the numbers keeping the units column aligned.

Example: 247×32

$$
\begin{array}{r}
1 \\
247 \\
\times 32 \\
\hline
4
\end{array}
\quad
\begin{array}{r}
1 \\
247 \\
\times 32 \\
\hline
94
\end{array}
\quad
\begin{array}{r}
1 \\
247 \\
\times 32 \\
\hline
494
\end{array}
\quad
\begin{array}{r}
2 \\
1 \\
247 \\
\times 32 \\
\hline
494 \\
1
\end{array}
\quad
\begin{array}{r}
12 \\
1 \\
247 \\
\times 32 \\
\hline
494 \\
41
\end{array}
\quad
\begin{array}{r}
12 \\
1 \\
247 \\
\times 32 \\
\hline
494 \\
7\ 41 \\
\hline
7,904
\end{array}
$$

Multiply the units or 1s columns ($2 \times 7 = 14$). Place the 4 below the 2 and carry the 1 to the next column. *Note that the 1 carried over is actually 10 carried to the next column, which is the 10s column.* Then multiply ($2 \times 4 = 8$) and add the 1 ($8 + 1 = 9$). *Since the 8 is in the 10s column, the answer is actually (80 + 10 = 90).* Place the 9 beside the 4. Multiply ($2 \times 2 = 4$). *The 2 is located in the 100s column. This is actually ($2 \times 200 = 400$).* Place the 4 beside the 9. Next, multiply each digit of 247 by 3 in 32. The answer will be brought down in the same manner except that one space will be skipped. *The reason for skipping the space is because the 3 in 32 is actually 30, not 3.* Multiply ($3 \times 7 = 21$). *The 21 is actually 210 ($30 \times 7 = 210$). Some people place a 0 under the number in the first column, in this case 4, just as a place holder.* Place the 1 below the 9 and carry the 2 to the next column. Multiply ($3 \times 4 = 12$). *This is actually ($30 \times 40 = 1200$).* Add the 2 from the first column ($12 + 2 = 14$). *This is actually (1200 + 200 = 1400).* Place the 4 beside the 1 and carry the 1 to the next column. *The 1 carried to the next column is actually 1,000 carried to the next column.* Multiply ($3 \times 2 = 6$). *This is actually ($30 \times 200 = 6,000$).* Add the 1 (6 + 1 = 7). *(6,000 + 1,000 = 7,000)*. Place the 7 beside the 4. The final step is to add the two sets of products together to obtain the total. The final answer is seven thousand, nine hundred and four.

PRACTICAL PROBLEMS

1. A panel board requires sixteen ½-inch holes, twenty-one ¼-inch holes, and eleven ⁵⁄₁₆-inch holes. Each hole requires a bolt with three washers and two nuts.

   a. Find the total number of washers needed for the ½-inch holes.   a. _____
   b. Find the total number of washers needed for the ¼-inch holes.   b. _____
   c. Find the total number of washers needed for the ⁵⁄₁₆-inch holes.   c. _____
   d. Find the total number of nuts needed for the ½-inch holes.   d. _____
   e. Find the total number of nuts needed for the ¼-inch holes.   e. _____
   f. Find the total number of nuts needed for the ⁵⁄₁₆-inch holes.   f. _____

2. A bearing on a large machine is tested over a period of 8 hours at a speed of 40,500 revolutions per hour. How many revolutions does the shaft turn in the bearing during the test period?   _____

3. Find the total amount of power in watts for the three motors shown. (One horsepower equals 746 watts.)   _____

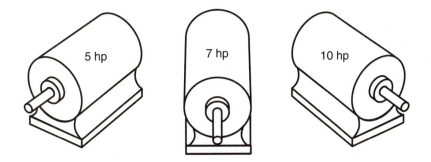

4. A very small magnet is wound with 97 layers with 215 turns per layer. How many turns of wire are on the coil?   _____

5. A coil requires 2,900 turns of number 14 wire. If each of 20 layers is wound with 143 turns, will the requirements for the coil be satisfied?   _____

6. A building uses the following size lamps: sixteen 50-watt, nine 15-watt, twelve 25-watt, six 75-watt, and four 100-watt. How many watts are consumed when all the lights in the building are burning?   _____

7. It is found that a certain electrical circuit having a total load of 2,800 watts in lamps must be reduced. Ten 200-watt lamps are replaced with ten 150-watt lamps; eight 100-watt lamps are replaced with eight 60-watt lamps. What is the total number of watts in connected lamps after the change is made?   _____

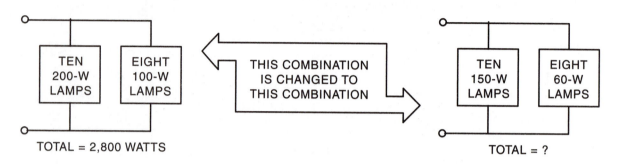

TOTAL = 2,800 WATTS   TOTAL = ?

8. Thirty-four steel boxes are used for a certain wiring job. In each box, five 1-inch holes are drilled. In twenty-three of the boxes, two 1¾-inch holes are drilled, and the remaining boxes have three 1½-inch holes drilled in them. How many holes are drilled in all the boxes?   _____

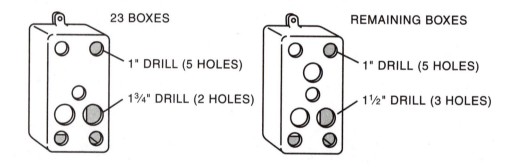

9. An electrical company has a payroll of 27 people. Seven people earn $18 per hour, eleven people earn $20 per hour, and nine people earn $16 per hour. If all employees work 40 hours during the week, what is the total payroll for one week?   _____

10. The product of current (amperes) and voltage (volts) equals power (watts). The total power of an electrical circuit is equal to the sum of the individual powers. Find the total power for the circuit shown.

_____

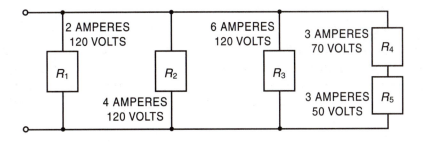

11. A large room contains 40 fluorescent lamps. Twenty-three of them are 40-watt lamps, and the remainder are 60-watt lamps. What is the total number of watts used by the lamps?

_____

12. A certain lighting job requires four incandescent fixtures, twelve direct fluorescent luminaires, and nine semidirect fluorescent luminaires. The incandescent fixtures cost $19 each, the direct luminaires cost $26 each, and the semidirect luminaires cost $31 each. Find the total cost of all the fixtures and luminaires.

_____

13. A nine-floor apartment building has an average of four electrical circuits for each apartment. There are six apartments on each floor and one additional apartment on the roof. Find the total number of electrical circuits in the apartment building.

_____

14. An electrician uses 763 feet of conduit on each floor of a five-story building. What is the total length of conduit used?

_____

 # Unit 4   DIVISION OF WHOLE NUMBERS

BASIC PRINCIPLES OF DIVISION OF WHOLE NUMBERS

In Unit 3 it was shown that multiplication is actually the process of adding a number to itself a certain number of times. Division is just the opposite or inverse. Division is actually the process of subtracting a smaller number from a larger number a certain number of times. The larger number, the number to be divided, is referred to as the *dividend.* The number used to indicate the number of times the dividend is to be divided is called the *divisor.* The answer is known as the *quotient*.

To begin the process of division, the dividend is placed inside the division bracket and the divisor is placed to the left of the dividend. The quotient is placed above the dividend.

$$\text{Divisor} \,\overline{)\,\text{Dividend}}^{\text{Quotient}}$$

Example: Divide 1,140 by 17.

$$
\begin{array}{r}
6 \\
17\overline{)\,1{,}140} \\
-1\,02 \\
\hline
12
\end{array}
\qquad
\begin{array}{r}
6 \\
17\overline{)\,1{,}140} \\
-1\,02 \\
\hline
120
\end{array}
\qquad
\begin{array}{r}
67 \\
17\overline{)\,1{,}140} \\
-1\,02 \\
\hline
120 \\
-119
\end{array}
\qquad
\begin{array}{r}
67\ \text{R1} \\
17\overline{)\,1{,}140} \\
-1\,02 \\
\hline
120 \\
-119 \\
\hline
1
\end{array}
$$

Place the number 1,140 under the division bracket and the number 17 to the left of it. The number 17 cannot be divided into a number that is smaller than itself. Therefore, 17 is divided into the number 114 first. Find what number multiplied by 17 will come the closest to 114 without going over 114. Six is that number (6 × 17 = 102). Place 6 in the quotient directly over 4 in the dividend. The number 102 is placed below 114 and subtracted from it. This leaves 12. Because 17 cannot be divided into 12, the next number of the dividend is brought down to the right of 12. When the zero is placed beside 12, 17 is then divided into 120. The nearest number that 17 can be multiplied by and not go over 120 is 7 (7 × 17 = 119). Place 7 in the quotient directly over 0 in the dividend, then place 119 below 120 and subtract from it. This leaves a remainder of 1. Since there are no more numbers in the dividend, the 1 is taken to the quotient and shown as R1, which means a *remainder* of 1.

Another way of understanding the process of division is to think of separating some number of objects into piles of fewer objects. Assume that an electrician finds 31 half-inch conduit clamps in a box. Now assume that each length of conduit will require 2 clamps. If he separates the 31 clamps into piles of 2 clamps each, how many piles of 2 clamps will he have? The answer is fifteen piles that contain 2 clamps and one pile that contains only 1 clamp; 31 divided by 2 is 15 with a remainder of 1.

PRACTICAL PROBLEMS

1. In a 184-foot run of BX cable, the staples are placed 4 feet apart. How many staples are used if one staple is placed at the beginning and one at the end of the run?   _____

2. An electrical contractor purchases 15 fittings of one type for $45 and 6 of another type for $36.

   a. Find the cost per fitting for those costing $45.   a. _____
   b. Find the cost per fitting for those costing $36.   b. _____

3. A total load of 25,620 watts is distributed equally over the 5 branch circuits shown. What is the average load per circuit in watts?   _____

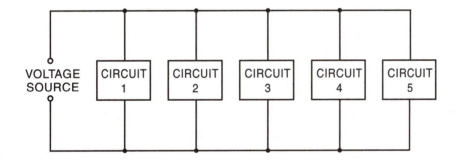

4. How many 250-foot rolls of BX cable are needed if a job requires a total of 5,250 feet?   _____

5. A hotel with 22 rooms on each of its seven floors has a total of 770 outlets. If each room has the same number of outlets, how many are there in each room?   _____

6. A certain wiring job has 28 outlets equally spaced over 351 feet. If one outlet is placed at the beginning and one at the end, what is the center-to-center distance between outlets?   _____

7. In a house where 35 outlets are installed, 735 feet of cable are used. What is the average number of feet of cable used per outlet?   _____

8. Twelve standard packages of conduit fittings are purchased. Their combined weight is 780 pounds. What is the weight per package?

_____

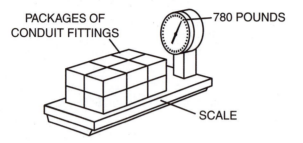

PACKAGES OF
CONDUIT FITTINGS

780 POUNDS

SCALE

9. Two electricians work a total of 640 hours on a job. Each works 8 hours per day, 5 days per week. How many weeks does each electrician work?

_____

10. Two thousand five hundred feet of BX cable are ordered. The cable is shipped in 250-foot coils. How many coils are shipped?

_____

11. A certain machine room uses 2,160 watts to supply 60-watt lamps for bench lighting. How many lamps are connected?

_____

12. A wiring job uses 1,232 feet of cable for 56 outlets. What is the average number of feet per outlet?

_____

13. The cost for the carton of staples shown is $36. What is the cost per standard package if the carton contains 12 standard packages?

_____

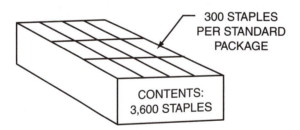

300 STAPLES
PER STANDARD
PACKAGE

CONTENTS:
3,600 STAPLES

14. On a 124-foot length of Romex cable, 32 staples are used. The staples are equally spaced. If one staple is placed at the beginning and one at the end of the cable, how far apart are the staples placed?

_____

15. A total load of 15,840 watts is distributed equally over 12 circuits. What is the load per circuit in watts? _____

16. Box A and box B each contain type C connectors. Box A contains 200 connectors and costs $30. Find the cost of box B, which contains 250 connectors. The unit price is the same for both boxes. _____

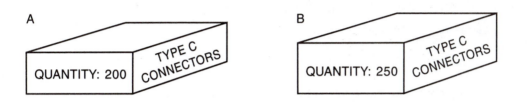

17. A school study hall is 90 feet by 90 feet. Two rows of lighting are placed in the ceiling. What is the center-to-center distance between the rows of lights if the rows are spaced with equal distances from the side walls and between the rows? _____

Unit 5 COMBINED OPERATIONS WITH WHOLE NUMBERS

This unit provides practical problems involving combined operations with addition, subtraction, multiplication, and division of whole numbers. There is a standard order that should be followed when dealing with combined operation.

1. Do all operations inside parentheses.

2. Solve any expressions that contain exponents or roots.

3. Multiply or divide from left to right.

4. Add or subtract from left to right.

Example: $(8 + 7) \times 4$ $(15) \times 4 =$ $15 \times 4 = 60$

Example: $(10 \times 2) \div 5$ $(20) \div 5$ $20 \div 5 = 4$

Example: $(18 + 10) - (6 + 12)\,(28) - (18)$ $28 - 18 = 10$

PRACTICAL PROBLEMS

1. In wiring eight houses, the electricians install 68, 87, 57, 74, 49, 101, 99, and 56 outlets. Find the total number of outlets that must be roughed-in.   _____

2. An electrician removes from stock, at different times, the following amounts of BX cable: 120 feet, 327 feet, 637 feet, 302 feet, 500 feet, 250 feet, 140 feet, 75 feet, and 789 feet. Find the total number of feet of BX cable taken from stock.   _____

3. An electrical supply house purchases solder in separate lots of 30 pounds, 120 pounds, 37 pounds, 125 pounds, 103 pounds, 33 pounds, 210 pounds, and 40 pounds. What is the total number of pounds of solder purchased?   _____

4. A school has twelve electrical circuits that carry 2,569 watts, 1,260 watts, 1,639 watts, 563 watts, 790 watts, 800 watts, 1,137 watts, 250 watts, 500 watts, 750 watts, 1,830 watts, and 2,462 watts. What is the total number of watts consumed when all these circuits are being used under their total loads?   _____

5. BX cable in the following amounts is used on an apartment house job: 250 feet, 71 feet, 39 feet, 110 feet, 75 feet, 87 feet, and 560 feet. What is the total amount of cable used on the job?   _____

6. The following number of BX staples are used during a given period: 28, 250, 38, 108, 92, 130, 25, 36, 97, 91, 65, and 40. Find the total number of BX staples used.   _____

7. An electrician takes out of stock 498 feet of BX cable on Monday, 103 feet on Tuesday, and 78 feet on Wednesday. On Friday, 27 feet of BX cable are returned to stock. How much BX cable is used?   _____

8. An inventory sheet shows a balance of 500 outlet boxes on January 1. On January 10, 127 outlet boxes are taken out of stock. On January 14, 61 outlet boxes are returned to stock. How many outlet boxes are left in stock after January 14?   _____

9. An electrical contractor charges $875 for a job. The materials cost $262. The cost of labor is $348, and the cost of transportation is $27. Find the profit.   _____

10. A purchase of 2,500 feet of number 14 double-braided, rubber-covered wire is made for a job. On November 1, 978 feet of this wire are used, and on November 3, 1,023 feet are used. How many feet of wire are left?   _____

11. A building contains seventy 100-watt lamps, thirty-eight 75-watt lamps, ten 60-watt lamps, and twenty 40-watt lamps. If all lamps are on at the same time, how many watts are used?   _____

12. An electrical contractor employs 16 people. Five people earn $15 per hour, four people earn $17 per hour, and the remaining people earn $16 per hour. What is the total hourly wage earned by all 16 people?

_____

13. A 7-floor apartment building has an average of 7 electrical circuits per apartment, and there are 8 apartments per floor. How many electrical circuits are there in the building?

_____

14. A wiring job requires 5,127 feet of cable. If the cable comes in 250-foot coils, how many coils of cable are required?

_____

15. A coil of wire is wound in 7 layers with 13 turns per layer. How many turns of wire are on the coil?

_____

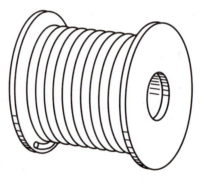

16. A wiring job requires 29 outlets to be spaced equally over 336 feet. One outlet is placed at the beginning of the 336 feet and one at the end. Find the center-to-center distance between the outlets.

_____

17. An order is placed for 16 coils of cable. The cable comes in 250-foot coils. How many feet of cable are received?

_____

18. Twenty standard cartons of octal boxes weigh a total of 1,100 pounds. Find the weight per carton.

_____

19. A box contains 315 half-inch conduit couplings and weighs 119 pounds. An identical box also contains half-inch conduit couplings and weighs 47 pounds. How many couplings are in the second box?

_____

20. Receptacle boxes are placed 12 feet apart. Holes are drilled in the wall studs 1 foot above the boxes to permit Romex wire to be run between them. Each receptacle box is 3 inches deep, and 6 inches of wire extend beyond the edge of a box. How many receptacle boxes can be wired with one box of Romex wire? (**Note:** A box of Romex wire contains 250 feet.) _____

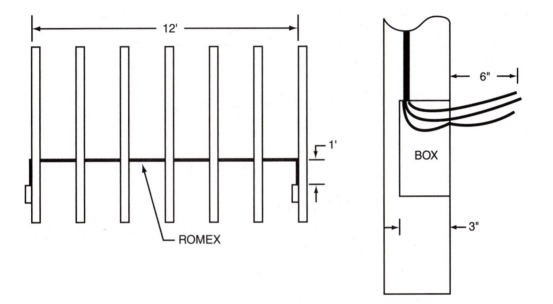

21. An electrician takes a job wiring 25 identical apartments. Each apartment contains 16 outlets that fit in a single-gang box and 6 single-pole switches that fit in a single-gang box. In addition to the single-gang switches, there are three two-gang switch boxes. Two of the two-gang switch boxes contain a single-pole switch and 3-way switch. The third two-gang box contains two single-pole switches.

   a. How many single-gang boxes are required to wire the 25 apartments?   a. _____
   b. How many two-gang boxes are required to wire the 25 apartments?   b. _____
   c. How many outlets are required to wire these apartments?   c. _____
   d. How many single-pole switches are required to wire these 25 apartments?   d. _____
   e. How many 3-way switches are required to wire these 25 apartments?   e. _____

Common Fractions

SECTION 2

Unit 6   ADDITION OF COMMON FRACTIONS

BASIC PRINCIPLES OF ADDITION OF COMMON FRACTIONS

Measurement often must be more precise than whole numbers allow. One method of indicating quantities that are smaller than a whole is that of using common fractions. There are two parts to a common fraction, the *numerator* (the number above the line) and the *denominator* (the number below the line).

The denominator indicates the number of equal parts the whole is divided into. The inch, for example, is often divided into 16 equal parts. The numerator indicates the number of parts used. If a measurement is $\frac{5}{16}$ inch, it indicates that an inch has been divided into 16 equal parts and the length corresponds to 5 of these 16 parts.

ADDING FRACTIONS

Before fractions can be added, all of their denominators must be the same. Then, only the numerators are added. $\frac{1}{16}$

Example: $\frac{1}{16} + \frac{3}{16} + \frac{5}{16}$

$$\frac{1}{16}$$

$$\frac{3}{16}$$

$$+\frac{5}{16}$$

$$\frac{9}{16}$$

FINDING A COMMON DENOMINATOR

If all the denominators are not the same, it is necessary to find a *common denominator.* A common denominator can be found by finding some number that all the denominators of the individual fractions will divide. To say a number *divides* another means no remainder is left in the division. In the following example, a common denominator could be 24, since the denominator of each fraction will divide 24. This is not to say that 24 is the only common denominator. Another common denominator is 48, for example. However, 24 is the lowest common denominator (LCD), and use of the LCD is preferred when adding fractions.

Example: $\frac{5}{12} + \frac{1}{4} + \frac{1}{6} + \frac{3}{24}$

To convert $\frac{5}{12}$ to an equivalent fraction with a denominator of 24, divide 24 by the original denominator of 12 (24 ÷ 12 = 2). Multiply the quotient by the numerator in the original fraction (5 × 2 = 10). The fraction $\frac{5}{12}$ is equal to $\frac{10}{24}$. The other fractions can be charged to equivalent fractions with a denominator of 24 in the same manner.

$$\frac{1}{4} = (24 \div 4 = 6)\ (1 \times 6 = 6)$$

$$\frac{1}{6} = (24 \div 6 = 4)\ (1 \times 4 = 4)$$

$$\frac{3}{24} = (24 \div 24 = 1)\ (3 \times 1 = 3)$$

Total

$$\frac{5}{12} = \frac{10}{24}$$
$$\frac{1}{4} = \frac{6}{24}$$
$$\frac{1}{6} = \frac{4}{24}$$
$$+\frac{3}{24} = \frac{3}{24}$$
$$\frac{23}{24}$$

Addition of these fractions is accomplished by changing each into an *equivalent fraction* with a denominator of 24. This is done by dividing the denominator of the fraction to be changed into the common denominator, and then multiplying the numerator by the answer. For example, $\frac{5}{12}$ is changed to $\frac{10}{24}$ because 24 ÷ 12 = 2 and 2 × 5 = 10. The fractions $\frac{5}{12}$ and $\frac{10}{24}$ are equal in value. To change $\frac{1}{4}$ into an equivalent fraction in 24ths, note that 24 ÷ 4 = 6 and 6 × 1 = 6. Therefore, the fraction $\frac{1}{4}$ has the same value as $\frac{6}{24}$. Once each fraction has been changed into an equivalent fraction in 24ths, the numerators are added together.

Finding a common denominator for the fractions in the preceding example was relatively simple because it was obvious that all denominators divided 24. There may be occasions, however, when a common denominator is not apparent. Assume that it is necessary to add $\frac{8}{17}$, $\frac{1}{5}$, and $\frac{3}{11}$. A number that can be divided by 17, 5, and 11 is not immediately apparent. A common denominator can always be found, however, by multiplying all the denominators together. This may not reveal the *lowest* common denominator, but it will produce a common denominator.

$$17 \times 5 \times 11 = 935$$

$$\frac{8}{17} \times \frac{5}{5} \times \frac{11}{11} = \frac{440}{935}$$

$$\frac{1}{5} \times \frac{17}{17} \times \frac{11}{11} = \frac{187}{935}$$

$$+\frac{3}{11} \times \frac{5}{5} \times \frac{17}{17} = \frac{255}{935}$$

$$\frac{882}{935}$$

REDUCING FRACTIONS TO LOWEST TERMS

It is common practice to reduce a fraction to its *lowest terms.* This is done by dividing both the numerator and denominator by the same number.

Example: Express $^{18}\!/_{48}$ in lowest terms.

$$\frac{18 \div 6}{48 \div 6} = \frac{3}{8}$$

Notice that 6 divides both 18 and 48. Therefore, the fractions $^3\!/_8$ and $^{18}\!/_{48}$ are equal in value. Another method that can be employed to reduce a fraction to its lowest terms is to convert both the numerator and denominator into their prime factors and then eliminate like factors. A prime number is any number that can be equally divided only by itself or 1. Examles of prime numbrs are 1, 2, 3, 5, 7, and 11.

$$\frac{18}{48} = \frac{2 \cdot 3 \cdot 3}{2 \cdot 3 \cdot 2 \cdot 2 \cdot 2}$$

$$\frac{18}{48} = \frac{\cancel{2} \cdot \cancel{3} \cdot 3}{\cancel{2} \cdot \cancel{3} \cdot 2 \cdot 2 \cdot 2} = \frac{3}{8}$$

Some fractions such as $^{23}\!/_{24}$ cannot be reduced because there is no number except 1 that will divide both 23 and 24. The fraction $^{23}\!/_{24}$ is in lowest terms.

The fractions considered so far had numerators smaller than their denominators. Such a fraction is called a *proper fraction.* When adding fractions, it is not uncommon for the answer to produce a fraction with a numerator greater than the denominator, such as $^{76}\!/_{24}$. This is called an *improper fraction,* which should be reduced to its lowest terms. The first step is to divide the numerator by the denominator. This will produce a whole number of some value and a remainder.

$$\begin{array}{r} 3 \text{ R4} \\ 24 \overline{)\ 76} \end{array}$$

The fraction then can be rewritten as:

$$3\frac{4}{24} = 3\frac{1}{6}$$

PRACTICAL PROBLEMS

1. Find the sum of each of the following.

    a. $\frac{5}{16}$, $\frac{3}{8}$, $\frac{3}{4}$, $\frac{7}{32}$

    b. $\frac{1}{16}$, $\frac{3}{4}$, $\frac{7}{8}$, $\frac{31}{32}$

    c. $\frac{3}{4}$, $\frac{8}{12}$, $\frac{5}{8}$, $\frac{5}{16}$

    d. $\frac{7}{10}$, $\frac{4}{5}$, $\frac{3}{4}$, $\frac{3}{10}$

    a. _____

    b. _____

    c. _____

    d. _____

Note: Use this diagram for problems 2 and 3.

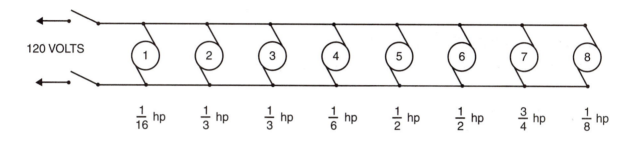

2. Eight motors, each having a rated horsepower as indicated on the diagram, are connected in a circuit. What is the total horsepower of motors 1 through 4 inclusive?

    _____

3. What is the total horsepower of motors 5 through 8 inclusive?

    _____

4. Eight resistances are connected in series as shown. The unit of measure for resistance is the ohm. The resistance in a series circuit is the sum of all resistances in that circuit. Find the total amount of resistance in the circuit.

    _____

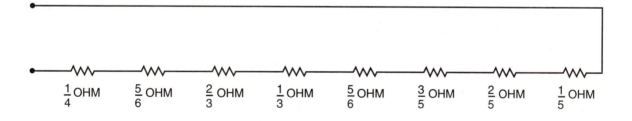

5. The insulation to be used in a certain slot is as follows: fish paper, $\frac{1}{64}$ inch; Tufflex, $\frac{1}{32}$ inch; varnished cambric, $\frac{1}{64}$ inch; and top stick, $\frac{1}{8}$ inch. What is the total thickness of the insulation?

    _____

6. A piece of carbon $\frac{31}{32}$ inch thick has a copper plating of $\frac{1}{64}$ inch on each side. What is the total thickness?

    _____

7. A motor is aligned using pieces of steel under the base. The thicknesses of the pieces are as follows: ¹³⁄₁₆ inch, ⁵⁄₆₄ inch, ³⁄₃₂ inch, and ⅛ inch. What is the total thickness of all these pieces?

_____

8. A copper contact is insulated from its support with one piece of mica ¹⁄₆₄ inch thick, two pieces of fiber that are each ⅛ inch thick, and ¹⁄₁₆ inch of pressboard. The copper is ¾ inch thick, and the support is ⅞ inch thick. What is the total thickness of copper, insulation, and support?

_____

9. A form is made to wind a coil. The allowances made for insulation are as follows: four layers of ¹⁄₆₄-inch tape, two pieces of ¹⁄₁₆-inch fish paper, and two pieces of ⅛-inch fiber. What is the total thickness allowed for insulation?

_____

10. On a theater repair job, it is found that a fixture has been hung from several pieces of wood spiked together. The bracket shown is made to take the place of the wood. The pieces of the bracket are all ⅝-inch thick. What is height (H) of the bracket?

_____

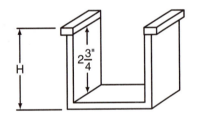

11. What is the total thickness of a wall having ⅞-inch finish siding, ⅞-inch rough siding, 3¾-inch studs, and ¹³⁄₁₆ inch of lath and plaster?

_____

12. Find the distance (A) for the bend of electrical conduit that is passing under the girder shown in the figure.

_____

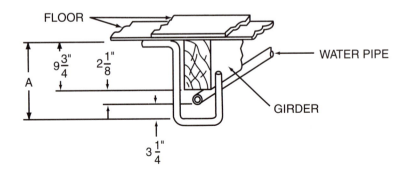

13. What space on a bolt will six washers or spacers of the following thicknesses occupy: ¹⁄₁₆ inch, ½ inch, ⅛ inch, ³⁄₃₂ inch, ⁵⁄₆₄ inch, and ³⁄₁₆ inch?

14. Find the shortest length of ½-inch conduit from which the following pieces can be cut: 7⅞ inches, 3¼ inches, 6½ inches, 12⅛ inches, and 24³⁄₁₆ inches. Allow ¹⁄₁₆ inch for each saw cut.

15. What is the shortest strip of fiber from which five pieces of the following lengths can be cut: ⅞ inch, 3⅛ inches, 2¹⁄₁₆ inches, 12¼ inches, and ¹⁄₁₆ inch? Allow ⅛ inch for each saw cut.

16. Find the total resistance (R_t) of a three-series connected resistors.

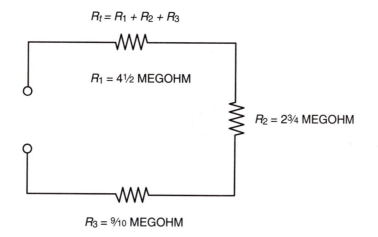

$$R_t = R_1 + R_2 + R_3$$

$R_1 = 4½$ MEGOHM

$R_2 = 2¾$ MEGOHM

$R_3 = \frac{9}{10}$ MEGOHM

17. Two-wire NM cable is used to connect four outlets. The distances between the outlets are 6⁹⁄₁₀ meters, 8⁹⁄₁₀ meters, 20½ meters, and 8⁹⁄₁₀ meters. Find the total length of wire between outlets.

Unit 7   SUBTRACTION OF COMMON FRACTIONS

BASIC PRINCIPLES OF SUBTRACTION OF COMMON FRACTIONS

Subtraction of common fractions is similar to addition of fractions. In both operations, it is first necessary to have common denominators. Once both fractions have a common denominator, the numerator of the smaller fraction can be subtracted from the numerator of the larger.

Example: ⅝ − ⁷⁄₁₆

$$\frac{5}{8} \times \frac{2}{2} = \frac{10}{16}$$

$$-\frac{7}{16} \times \frac{1}{1} = \frac{7}{16}$$

$$\frac{3}{16}$$

In this example, ⁷⁄₁₆ is subtracted from ⅝. The lowest common denominator for these two fractions is 16. The fraction ⅝ is changed to ¹⁰⁄₁₆: (16 ÷ 8 = 2; 2 × 5 = 10). The fraction ⁷⁄₁₆ is not changed. The numerator of the second fraction is then subtracted from that of the first (10 − 7 = 3). The answer is ³⁄₁₆.

PRACTICAL PROBLEMS

1. Two pieces of rigid conduit (electrical pipe) are measured. Each piece has an approximate outside diameter of 1¼ inch, and the thickness of the wall is approximately ⅛ inch. Find the approximate inside diameter of each piece of conduit.   _____

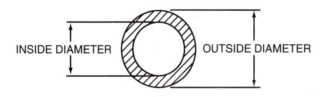

INSIDE DIAMETER   OUTSIDE DIAMETER

2. The slot shown is $\frac{5}{16}$ inch wide and ½ inch deep. The piece of fiber is $\frac{5}{64}$ inch thick. How deep is the space left for wires?   _____

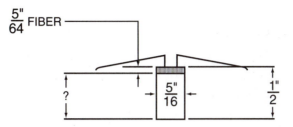

3. A washer ⅛ inch thick and a washer $\frac{3}{32}$ inch thick are placed on a bolt that is 1½ inches long under the head. What is the distance (A) left between the two washers after the ½-inch thick nut is screwed to its full thickness?   _____

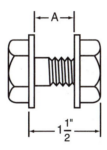

4. A 1½-inch bolt is inserted through a block $\frac{11}{16}$ inch thick with a ⅛-inch washer and a ½-inch nut. How much of the bolt will extend beyond the nut when drawn up tightly?   _____

5. The length of threading on a BX box connector is approximately $\frac{7}{16}$ inch. The locknut is ⅛ inch thick, and the metal into which this connector is inserted is $\frac{1}{32}$ inch. How much threading is left for a threaded bushing?   _____

6. The diameter of a steel shaft, through wear, is reduced $\frac{5}{1,000}$ inch. The shaft measured $\frac{875}{1,000}$ inch originally. The electrician removes another $\frac{5}{1,000}$ inch and makes a new bearing to fit the shaft. What is the new size of the shaft?   _____

7. A motor base must be blocked up 6⅛ inches. Two blocks are used. If one block is 3¾ inches thick, how thick is the other?   _____

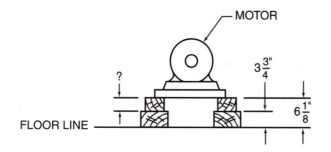

8. A cable measures $\frac{500"}{1,000}$ inch outside diameter and has $\frac{12"}{1,000}$ inch of cotton and $\frac{50"}{1,000}$ inch of rubber insulation. Determine the diameter of the copper conductor (wire).   _____

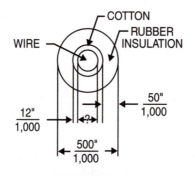

9. A ²⁹⁄₃₂-inch hole must be enlarged to ⁵⁹⁄₆₄ inch to insert a bushing. How much wider than the original hole will this be?   _____

10. A motor commutator 3⅛ inches in diameter is turned down to remove a flat spot. If ³⁄₆₄ inch is removed from the surface, find the finished diameter.   _____

Unit 8 MULTIPLICATION OF COMMON FRACTIONS

BASIC PRINCIPLES OF MULTIPLICATION OF COMMON FRACTIONS

Fractions are multiplied by simply multiplying the numerators together and the denominators together.

Example: Multiply ⅝ by ¼.

$$\frac{5}{8} \times \frac{1}{4} = \frac{5}{32}$$

When fractions are multiplied, it is often possible to simplify the problem by *cross reduction.* If a number can be found that will divide into both the numerator of one fraction and the denominator of the other, the problem can be made simpler. The answers should always be reduced to lowest terms.

Example: ⅘ × ¹⁵⁄₃₂

$$\frac{\overset{1}{\cancel{4}}}{\underset{1}{\cancel{5}}} \times \frac{\overset{3}{\cancel{15}}}{\underset{8}{\cancel{32}}} = \frac{3}{8}$$

When a fraction is to be multiplied by a whole number, the whole number is first changed into a fraction. This is done by placing the whole number over 1.

Example: 3 × ⁷⁄₁₂

$$\frac{3}{1} \times \frac{7}{12} = \frac{1}{1} \times \frac{7}{4} = 1\frac{3}{4}$$

There are also times when it is necessary to change a *mixed* number into an *improper fraction.* A mixed number is a number that contains both a whole number and a fraction, such as 3½. To change a mixed number into an improper fraction, multiply the whole number by the denominator and then add the numerator to the result. Place the answer in the numerator of the improper fraction. Leave the denominator as it was.

Example: Change 3½ into an improper fraction

$$3\frac{1}{2} = \frac{3 \times 2 + 1}{2} \times \frac{6 + 1}{2} = \frac{7}{2}$$

11. The motor sleeve bearing shown in the figure has an outside diameter of 3¹⁄₁₆ inches. The thickness of the wall is ¹⁹⁄₆₄ inch. Find the inside diameter.

_____

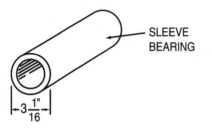

SLEEVE
BEARING

$-3\frac{1"}{16}-$

12. A motor brush is 1⅞ inches long. How long is it after ⁴⁹⁄₆₄ inch wears away?

_____

13. Kirchhoff's law is used for circuits when the conductors forming a part of a network carrying currents meet at one point. Kirchhoff's law states that the current entering the electrical connection is equal to the sum of the currents leaving the connection. In the circuit shown, find the current, I_2.

_____

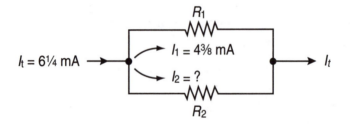

R_1

$I_t = 6\frac{1}{4}$ mA

$I_1 = 4\frac{3}{8}$ mA

$I_2 = ?$

I_t

R_2

14. An electrician uses 46¾ feet of push-back wire from a 200-foot roll. How many feet of wire are left on the roll?

_____

Example: $3\frac{5}{8} \times 4\frac{1}{4}$

The first step is to change both mixed numbers into improper fractions. The improper fractions can then be multiplied.

$$3\frac{5}{8} = \frac{29}{8}$$

$$4\frac{1}{4} = \frac{17}{4}$$

$$\frac{29}{8} \times \frac{17}{4} = \frac{493}{32}$$

$$\frac{493}{32} = 15\frac{13}{32}$$

PRACTICAL PROBLEMS

1. The ceiling outlets for the circuit weigh $\frac{11}{16}$ pound each. Find the total weight for all the ceiling outlets. _____

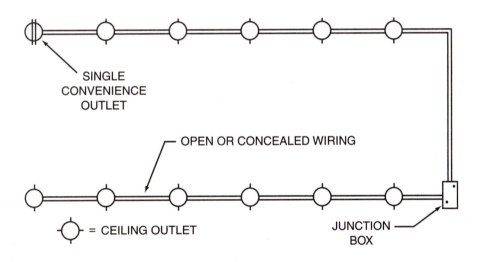

SINGLE CONVENIENCE OUTLET

OPEN OR CONCEALED WIRING

$\bigcirc$ = CEILING OUTLET

JUNCTION BOX

2. If an electrician works $4\frac{3}{4}$ hours a day on a job at \$16 per hour, how much money does he earn in 5 days? _____

3. A wiring job calls for thirty-two pieces of $\frac{1}{2}$-inch conduit $7\frac{1}{2}$ feet long, eight pieces $13\frac{1}{4}$ inches long, three pieces $7\frac{3}{4}$ inches long, and six pieces $9\frac{5}{8}$ inches long. What is the total number of feet of $\frac{1}{2}$-inch conduit required for the job? _____

4. A department in a factory has three ¾-horsepower motors, five ¼-horsepower motors, six 3⅓-horsepower motors, and eight 7½-horsepower motors. What is the total connected motor load in horsepower rating?   _____

5. A wiring job requires BX cable in the following lengths: eight pieces 23½ feet each, seven pieces 18½ inches each, twelve pieces 24½ inches each, and twenty-five pieces 19½ inches each. How many feet of BX cable are needed?   _____

6. In estimating a job, it is decided that it should take 13 people 3½ hours each to do part of the work, and 7 people 6¾ hours each to do the remainder of the job. Determine the total number of hours estimated for the job.   _____

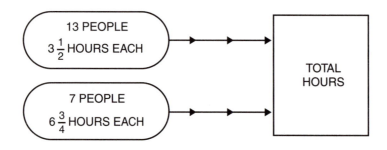

7. For the circuit shown, determine the sum of all the energy used in kilowatt-hours if all units are on for 3¾ hours. The power shown for each unit is in kilowatts.   _____

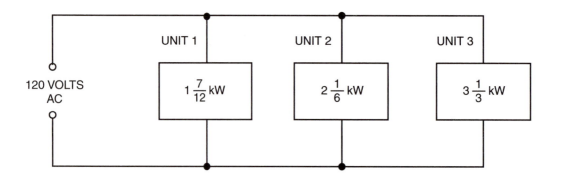

8. A standard package of 3-inch molding junction boxes weighs 6¼ pounds. If 16 packages are purchased, find the total weight.   _____

9. A storage battery contains 1¼ kilograms of electrolyte. One-third of the electrolyte is acid. How many kilograms of acid are in the battery cells?   _____

10. A 75-watt, 120-volt light uses ⅝ ampere. How many amperes are used by the 5 lights connected in parallel?   _____

 # Unit 9 DIVISION OF COMMON FRACTIONS

BASIC PRINCIPLES OF DIVISION OF COMMON FRACTIONS

Common fractions can be divided in a manner similar to that used to multiply common fractions. When dividing common fractions, it is necessary to invert the divisor (the second fraction) and multiply. The same rules that are used for the multiplication of fractions can then be followed. All answers should be reduced to lowest terms.

Example: Divide ¾ by ⅝.

$$\frac{3}{4} \div \frac{5}{8} = \frac{3}{\overset{}{\underset{1}{4}}} \times \frac{\overset{2}{8}}{5} = \frac{6}{5} = 1\frac{1}{5}$$

Example: When a whole number is to be divided by a fraction, the whole number is placed over one and then treated as a fraction.

$$8 \div \frac{1}{4} = \frac{8}{1} \div \frac{1}{4} = \frac{8}{1} \times \frac{4}{1} = \frac{32}{1} = 32$$

Example: It is sometimes possible to reduce fractions by cross division. The top of one fraction and the bottom of the other may be reduced by finding some number that will divide into each of them an even number of times. In this example, ⁵⁄₁₂ is divided by 10. The 10 is placed over 1 and then invertred to become ¹⁄₁₀. The 5 of ⁵⁄₁₂ and the 10 of ¹⁄₁₀ can both be divided by 5. Five will divided into 5 one time and into 10 two times. The equivalent problem becomes ¹⁄₁₂ times ½. The answer is ¹⁄₂₄.

$$\frac{5}{12} \div 10 = \frac{5}{12} \div \frac{10}{1} = \frac{5}{12} \times \frac{1}{10} = \frac{\overset{1}{5}}{12} \times \frac{1}{\underset{2}{10}} = \frac{1}{24}$$

PRACTICAL PROBLEMS

Note: No waste allowance is made for cutting unless indicated in the problem.

1. How many ⅞-inch lengths can be cut from a 7-inch fiber strip?   _____

2. How many pieces ¹⁵⁄₁₆ inch long can be cut from a strip of sheet metal 13⅛ inches long?   _____

3. If 12⅞ watts are distributed equally over each of the resistors shown, find the average number of watts per resistor.   _____

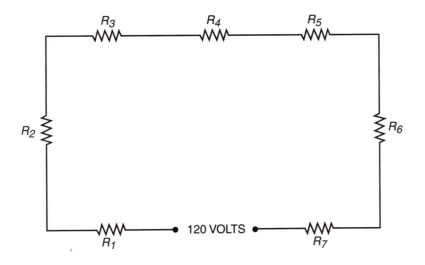

4. A certain circuit is properly fused for a 5-horsepower motor. If the motor is to be replaced by several ⅝-horsepower motors, how many motors will the fuses be able to carry?   _____

5. A 20-foot length of underground cable is cut into 6¾-inch pieces. How many 6¾-inch pieces are made?   _____

6. How many whole wedges 3⅞ inches long can be made from 20 wedge strips, each 3 feet long?   _____

7. Two electricians work on an electrical job in an apartment building. Both work five days per week. Over a two-week period, how many times longer than electrician A does electrician B work, based on the hours shown in the graph?   _____

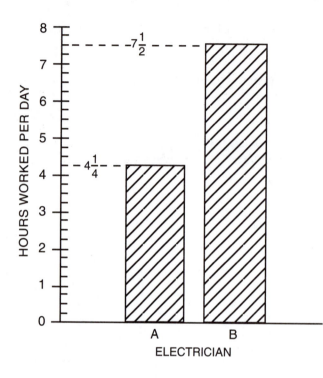

8. How many 1¾-inch-long machine bolt blanks can be cut from a 5-foot length of stock? Allow ⁷⁄₃₂ inch for waste on each blank.   _____

9. If 7¾ dozen ½-inch connector fittings cost $31, what is the cost per dozen?   _____

10. One meter of shielded microphone cable weighs ⁷⁵⁄₁,₀₀₀ kilogram. What is the length of cable in a coil that weighs 5½ kilograms?   _____

Unit 10 COMBINED OPERATIONS WITH COMMON FRACTIONS

This unit provides practical problems involving combined operations of addition, subtraction, multiplication, and division of common fractions.

PRACTICAL PROBLEMS

1. Eight motors are connected in a circuit. The horsepower ratings are ⅛, ¾, 1/16, ½, ⅓, ½, ⅓, and ¼. What is the total horsepower of the circuit? _____

2. The insulation to be used in a certain slot is as follows: fish paper, 1/64 inch; Tufflex, 1/16 inch; varnished cambric, 1/32 inch; and top stick, ⅛ inch. What is the total thickness of all the insulation? _____

3. A copper contact is insulated from its support with one piece of mica ⅛ inch thick, two pieces of fiber each 1/16 inch thick, and ⅛ inch of pressboard; the copper is 1/32 inch thick, and the support is ¼ inch thick. What is the total thickness of copper, insulation, and support? _____

4. In a motor alignment, the following pieces of steel are used under the base: one piece 5/16 inch thick, one piece ⅝ inch thick, one piece 5/32 inch thick, and one piece 5/64 inch thick. What is the total thickness of all these pieces? _____

5. A form is made to wind a coil. Allowance is made for insulation as follows: four layers of 1/32 inch tape, two pieces of ⅛-inch fish paper, and two pieces of ¼ inch fiber. What is the total thickness allowed for insulation? _____

6. What space on a bolt will six washers or spacers of the following thicknesses occupy: 1/64 inch, 1/32 inch, 1/64 inch, 1/32 inch, ⅛ inch, and 3/16 inch? _____

7. What is the shortest length of ½-inch conduit from which the following pieces can be cut: 3⅞ inch, 5½ inch, 7¾ inch, 9⅛ inch, and ⅜ inch? Allow 1/64 inch for saw cuts. _____

8. What is the shortest strip of fiber from which five pieces of the following lengths can be cut: 9½ inches, 10 11/16 inches, 6 9/16 inches, 7½ inches, and 1⅝ inches? No allowance is made for waste. _____

9. A slot in a piece of copper is too wide for a wire to be held tightly. The slot is ½ inch in width and the wire is ⅜ inch in diameter. How much wider is the slot than the wire? _____

10. The diameter of a steel shaft, through wear, is reduced $^6/_{1,000}$ inch. The shaft measures $^{875}/_{1,000}$ inch originally. The electrician removes another $^9/_{1,000}$ inch and makes a new bearing to fit the shaft. What is the size of the shaft after the work is completed?   _____

11. A cable measures $^{625}/_{1,000}$-inch outside diameter and has $^{16}/_{1,000}$ inch of cotton and $^{62}/_{1,000}$ inch of rubber insulation. What is the diameter of the copper conductor (wire)?   _____

12. A $^{28}/_{32}$-inch hole must be enlarged to $^{58}/_{64}$ inch to insert a bushing. How much wider than the original hole will the new hole be?   _____

13. A wiring job calls for 36 pieces of ½-inch conduit, each 6½ feet long; 9 pieces, each 11 inches long; 4 pieces, each 6 inches long; and 5 pieces, each 18 inches long. What is the total number of feet of ½-inch conduit required for the job?   _____

14. The ceiling outlets on a residential job weigh an average of ⅝ pound each, and 12 are required. Find the total weight of the outlets.   _____

15. A wiring job requires 2-conductor BX cable in the following lengths: 5 pieces, each 3½ feet long; 8 pieces, each 18 inches long; and 6 pieces, each 30 inches long. How many feet of 2-conductor BX are used?   _____

16. A department in a factory has three 7½-horsepower motors; four 1½-horsepower motors; two ¼-horsepower motors; and three ½-horsepower motors. What is the total connected motor load in horsepower rating?   _____

17. If an electrician works 7½ hours a day on a certain job at $25.60 an hour, what is the pay for a 5-day workweek?   _____

18. How many whole wedges 6¾ inches long can be made from 2 wedge strips, each 5¼ feet long?   _____

19. If 7⅓ yards of varnished cambric (insulation) cost $22, what is its cost per yard?   _____

20. How many 6¾-inch lengths can be cut from a 20-foot length of conduit?   _____

21. If a motor has a speed of 1,627¾ revolutions per second, how many revolutions will it make in $^{11}/_{15}$ second?   _____

22. Two electricians are assigned work on a remote-control wiring job. One electrician works 7½ hours each day, and the other electrician works 1½ hours each day. If they both work for 5 days, how many times longer does the first electrician work than the second electrician?   _____

23. A box of motor brushes costs 7½ dollars. If the price of one brush is ¼ dollar, how many brushes are in the box?   _____

24. If 7½ kilowatts of power are distributed equally over 5 resistors, what is the average number of kilowatts per resistor?   _____

25. A reel of annunciator wire is purchased for 10½ cents per foot, and the total cost is $525. How many feet of wire are on the reel?   _____

26. A 10 foot length of electrical metallic tubing is cut into 1½-foot lengths. How many whole pieces are made?   _____

27. An average of 3½ wire nuts is used in each of 14 octal boxes in a home. How many wire nuts are used in all?   _____

28. Two electricians work 6½ hours each per day for 5 days. Find the total number of hours worked.   _____

Decimal Fractions

Unit 11   ADDITION OF DECIMAL FRACTIONS

BASIC PRINCIPLES OF ADDITION OF DECIMAL FRACTIONS

One of the advantages of decimal fractions as compared to common fractions is that when decimal fractions are added or subtracted, it is not necessary to find a common denominator. Adding decimal fractions is accomplished by placing the fractions in a column with all the decimal points aligned. The columns are then added in the same manner as whole numbers are added.

It should be noted that each column to the left of the decimal increases in value by a factor of ten after the 1s or units column. Each column to the right of the decimal point decreases by a factor of ten. Some columns and values are shown below.

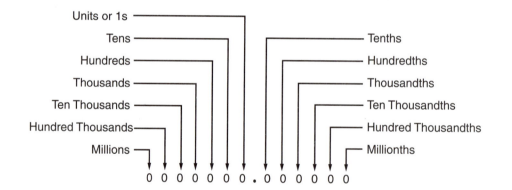

Example: 4.0563 + 0.98 + 14.0008 + 0.4005 + 10.33

$$
\begin{array}{r}
4.0563 \\
0.9800 \\
14.0008 \\
0.4005 \\
+ \ 10.3300 \\
\hline
29.7676
\end{array}
$$

PRACTICAL PROBLEMS

1. An electrical contractor purchases the following: 750 feet of 2-conductor BX cable for $205.50; 250 feet of 3-conductor BX cable for $48.38; and 100 feet of ½-inch electrical conduit (pipe) for $32.38. What is the total bill for these items?

    _____

2. What is the total thickness of these shims taken from a bearing: 0.007 inch, 0.125 inch, 0.140 inch, 0.187 inch, and 0.004 inch?

    _____

3. Materials and labor required to repair an electrical switch are as follows: one shallow box at $1.50; one ⅜-inch hickey at $0.65; one flush toggle switch at $1.98; and labor at $55.92. What is the total cost to repair an electrical outlet?

    _____

4. The total current (amperes) is equal to the sum of the individual currents. Find the total current in the figure shown.

    _____

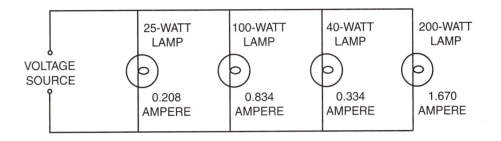

5. What is the cost of winding an alternator if it takes 1 pound of number 16 wire at $3.98; 1 pint of insulating varnish at $1.12; 2 ounces of insulation at $0.49; 2 ounces of gray fiber at $0.57; ½ yard of varnished cambric at $0.38; and labor costing $76.40?

    _____

6. The thicknesses of insulation between the copper conductor and the iron core of a motor armature are 0.002 inch of enameled insulation, 0.006 inch of cotton insulation, 0.010 inch of slot insulation, and 0.008 inch of varnished cambric insulation. Find the total thickness of insulation.

    _____

7. The costs of rewinding a motor are as follows: number 17 wire at a cost of $1.43; number 28 wire at a cost of $0.97; ½ pint of armalac at $1.89; and labor costing $118.60. Find the total cost of rewinding the motor.

    _____

8. A contractor requires the following for a job: BX cable, $32.65; conduit pipe, $12.56; toggle switches, $24.45; octal boxes, $7.50. What is the total cost of these materials?

    _____

9. The total current is measured by the ammeter (A). The total current in amperes is distributed in the circuit as shown. Find the total current reading on the ammeter.

_____

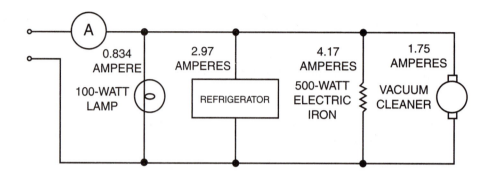

10. The values of pieces of electrical equipment are as follows: one 0-150 voltmeter costing $45.50; one 0-10-kilowatt meter costing $165.50; one 0-30 ammeter costing $35.20; one 0-300 voltmeter costing $42.75; one 0-200 millivoltmeter costing $65.38; one 50-ampere shunt costing $9.85; and three 0-5 ammeters costing $45.60 each. Find the total value of these pieces.

_____

Note: Use this diagram for problems 11–15.

_____

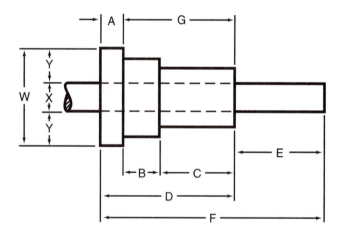

11. If A = 0.375 inch, B = 0.875 inch, and C = 1.062 inches, find D.

_____

12. If A = 0.3125 inch, B = 0.500 inch, and C = 1.062 inches, find D.

_____

13. If A = 0.75 inch, B = 0.500 inch, C = 2.625 inches, and E = 3.125 inches, find F.

          —————

14. If A = 0.375 inch, B = 0.923 inch, and C = 1.0992 inches, find G.

          —————

15. If X = 0.9375 inch and Y = 0.875 inch, find W.

          —————

16. The inside diameter of a motor shaft bearing is 0.0075 centimeter wider than the shaft. The shaft diameter is 4.75 centimeters. What is the inside diameter of the bearing?

          —————

17. The inside diameter of the conduit shown is 4.035 centimeters. It is made of 0.3-centimeter-thick steel. Find the outside diameter.

          —————

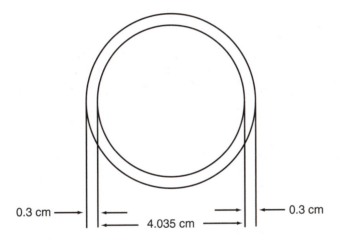

0.3 cm →　　　← 0.3 cm

4.035 cm

18. An electrician uses the following amounts of 1-inch EMT conduit: 22.5 feet, 8 feet, 30.25 feet, 22.5 feet, and 816.75 feet. Find the total amount of EMT used.

          —————

Unit 12 SUBTRACTION OF DECIMAL FRACTIONS

BASIC PRINCIPLES OF SUBTRACTION OF DECIMAL FRACTIONS

When decimal fractions are subtracted, the smaller number is placed below the larger number with the decimal points aligned. The procedure is then the same as that used in the subtraction of whole numbers. The decimal point in the answer is aligned with those above it.

Example: 12.349 − 8.907

$$
\begin{array}{r}
12.349 \\
-\ 8.907 \\
\hline
3.442
\end{array}
$$

PRACTICAL PROBLEMS

1. A power plant generates 6,336.71 kilowatt-hours on Monday and 5,269.36 kilowatt-hours on Tuesday. How much less does the plant generate on Tuesday as compared to Monday?

    _____

2. The voltage at the terminals A and B in the diagram of the electric generator shown is 219.7 volts, and the voltage drop (loss in the line) is 3.96 volts. What is the voltage at the end of the line at points C and D?

    _____

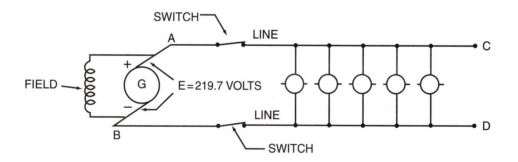

3. The total weight of a spool of number 16 wire is 11.01 pounds. The weight of the spool alone is 1.89 pounds. Find the weight of the wire only.

    _____

4. The cold resistance of an armature measures 0.096 ohm. After running for 5 hours, the resistance is 1.25 ohms. What is the difference in the resistance?   _____

5. A pin insulator measures 1.312 inches in diameter at the small end and 1.9375 inches at the large end. What is the difference in diameters between the small end and the large end?   _____

6. The field current of a motor is 1.12 amperes at no load. However, when the full load is applied, the current increases to 1.87 amperes. What is the difference in amperes from no load to full load?   _____

7. A bearing bushing has an outside diameter of 3.9375 inches and a wall thickness of 0.28125 inch. Find the inside diameter of the bushing.   _____

8. A special resistance wire has a diameter of 0.037 inch. The next smaller size is 0.0345 inch in diameter. Find the difference in the diameters.   _____

Note: Use this diagram for problems 9–13.   _____

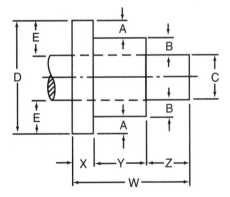

9. If D = 1.875 inches and E = 0.3125 inch, find C.   _____

10. If D = 3.1875 inches and E = 0.5625 inch, find C.   _____

11. Find C if A = 0.375 inch, B = 0.25 inch, and D = 2.3125 inches.   _____

12. If W = 3.000 inches, Y = 1.750 inches, and Z = 0.875 inch, find X.   _____

13. Find Z if X = 0.250 inch, Y = 2.0625 inches, and W = 3.375 inches.   _____

Note: Use this table for problems 14–17.

SOLID BARE COPPER CONDUCTORS

AMERICAN WIRE GAUGE SIZE NUMBER	WIRE DIAMETER IN INCHES
10	0.10190
11	0.09074
12	0.08081
13	0.07196
14	0.06408
15	0.05707
16	0.05082

14. How much larger is the diameter of number 11 wire than that of number 16 wire?

_____

15. Is number 10 wire larger or smaller than number 12 wire?

_____

16. What is the difference in diameters between number 13 wire and number 15 wire?

_____

17. Sometimes in the manufacture of wire, the wire is slightly larger or smaller than intended. If a wire is measured and found to be 0.079 inch in diameter, for what standard wire size is it intended?

_____

18. A shim is made 0.0025 inch thick. The thickness should be 0.00225 inch. Find the difference in the thicknesses.

_____

19. A motor shaft 1 inch in diameter is worn down to 0.9965 inch. How much is the diameter reduced by wear?

_____

20. The total current in the diagram divides at point A. How much current exists at point B?

_____

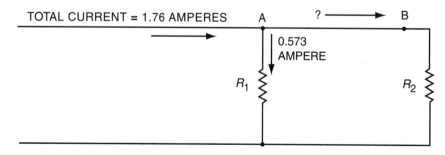

TOTAL CURRENT = 1.76 AMPERES   A   ? ———→ B

0.573 AMPERE

R_1 R_2

 # Unit 13 MULTIPLICATION OF DECIMAL FRACTIONS

BASIC PRINCIPLES OF MULTIPLICATION OF DECIMAL FRACTIONS

The multiplication of decimal fractions is the same basic procedure as the multiplication of whole numbers. When decimal fractions are multiplied, the number of places to the right of the decimal point in both numbers must be counted. The same number of places must appear to the right of the decimal point in the answer.

Example: 8.650 × 3.5

$$
\begin{array}{r}
8.650 \\
\times\ 3.5 \\
\hline
4\,3250 \\
25\,950 \\
\hline
30.2750
\end{array}
$$

8.650 (Three decimal places)
× 3.5 (One decimal place)

30.2750 (Four decimal places)

PRACTICAL PROBLEMS

1. How much does it cost to operate a heater that uses 0.45 kilowatt of power if it is kept on for one hour? The cost is $0.38 per kilowatt-hour. _____

2. Find the total weight of a 52-gallon steel drum and its contents if it contains 52 gallons of varnish weighing 8.16 pounds per gallon. The drum alone weighs 37.5 pounds. _____

3. The circumference of a circle equals the diameter multiplied by 3.1416. Find the circumference to the nearer hundredth of a pulley that has a diameter of 18.5 inches. _____

4. It is necessary to band an armature with 18 turns of steel wire per band. If the armature has a diameter of 20 inches, how many feet of wire are necessary per band, allowing 1.5 feet per band for waste and tightening? _____

5. A commutator has a diameter of 2.375 inches. What is the circumference to the nearer thousandth of the commutator? _____

6. If one cubic inch of copper weighs 0.314 pound, find the weight of a piece of copper that contains 2.25 cubic inches. _____

7. What is the weight of one gallon of sulphuric acid if 1 cubic inch weighs 0.0665 pound? (231 cubic inches = 1 gallon)   _____

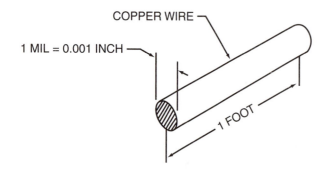

COPPER WIRE

1 MIL = 0.001 INCH

1 FOOT

8. What is the resistance of a piece of copper wire that has a size of 2.5 mil-feet if 1 mil-foot has a resistance of 10.4 ohms?   _____

9. An electric meter registers 8,749 kilowatt-hours on July 1 and 8,930 kilowatt-hours on August 1. The difference between these values is the total amount of energy used during this period. At $0.35 per kilowatt-hour, find the cost of the energy used.   _____

10. What is the cost of 1,250 feet of cable at $23.25 per one hundred feet?   _____

11. What is the cost of 1,750 feet of BX cable if it sells for $0.6823 per foot?   _____

12. What is the tax bill for a contractor if his shop has an assessed valuation of $35,000, and the tax rate is $27.50 per thousand dollars of assessed valuation?   _____

13. A truck is known to consume about 0.12 gallon of gas per mile. If it is driven 18,500 miles and gas costs $1.869 per gallon, what is the cost for gas?   _____

14. The circumference of a pulley equals the diameter multiplied by 3.1416. Find the difference in circumferences between the two pulleys shown. _____

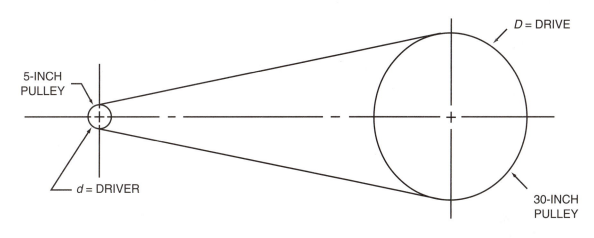

15. The drive pulley for the conveyer belt shown turns at the rate of 20 revolutions per minute. How many feet does the conveyer belt travel in one minute? (The distance for one revolution equals the circumference of the pulley.) _____

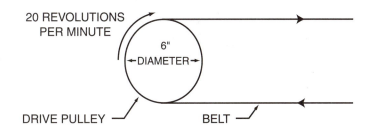

16. Six hundred twenty-five watertight screw connectors are purchased for $0.1725 each. Find the total cost. _____

17. A dry cell delivers about 1.375 volts. What voltage will 27 dry cells deliver if connected in series? (When connected in series, the voltage equals the sum of the voltages of the cells.) _____

18. A power company charges a city $1.2167 per street lamp per month. There are 1,125 street lamps supplied with current. What is the city's power bill per year for the street lamps? _____

 # Unit 14 DIVISION OF
DECIMAL FRACTIONS

BASIC PRINCIPLES OF DIVISION OF DECIMAL FRACTIONS

When decimal fractions are divided, the divisor is placed to the left of the dividend in the same manner as in the division of whole numbers. When dividing decimal fractions, however, the divisor must be a whole number and not a fraction. The divisor can be made a whole number by moving the decimal point all the way to the right of the number. When this is done, the decimal point of the dividend must be moved the same number of places to the right. The decimal point of the dividend is then placed directly above the division bracket. The numbers are then divided in the same manner as whole numbers.

Example: 19.44 ÷ 3.6

$$
\begin{array}{r}
5.4 \\
36.\overline{)\,194.4} \\
-180 \\
\hline
144 \\
-144 \\
\hline
\end{array}
$$

PRACTICAL PROBLEMS

1. If it costs $1,877 to construct ⅓ of a mile (5,280 feet = 1 mile) of an underground transmission system, what is the average cost of this job per foot? Express the answer to the nearest tenth of a cent.   _____

2. The lamps in the figure require a total current of 7.76 amperes. How many amperes will 2 lamps require? (All the lamps are of the same type, and each requires the same number of amperes.)   _____

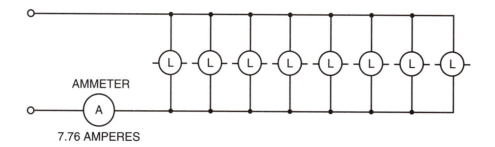

3. A certain size wire has a resistance of 2.56 ohms per one thousand feet. What is the resistance of 1 foot? _____

4. An electrician receives $787.50 for a 35-hour week. What is his rate of pay per hour? _____

5. If porcelain insulating tubes cost $10.56 per thousand, what is the loss to a school over a period of 27 school days if an average of 3 tubes are broken each day? Express the answer to the nearer cent. _____

6. A contractor pays an electrician $910 per week and a helper $560 per week. If the workweek consists of 35 hours, find the cost per hour for labor on a job using two electricians and two helpers. _____

7. The holes in the panel board shown are equally spaced across the board. Each hole is 1.25 inches in diameter, and the distance from each edge of the panel to the nearest hole is equal to the distance between holes. What is the length of the spaces between the holes? _____

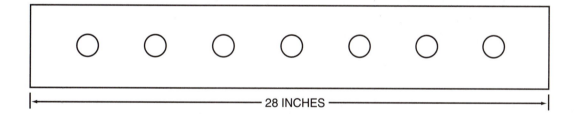

|←———————————— 28 INCHES ————————————→|

8. The weight of a spool of wire is 27.11 pounds. The spool weighs 3.2 pounds. Find the cost of the wire per pound if the cost of the spool of wire is $74.44. Express the answer to the nearer tenth of a cent. _____

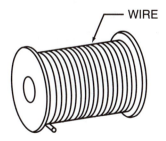
WIRE

9.   The power for a circuit is 1,265.75 watts. Seven equal units use this amount of power. Find the number of watts to the nearer hundredth used per unit.   _____

10. The lighting set shown has the same number of volts across each lamp. The sum of all lamp voltages is equal to the total number of volts. Find the number of volts across each lamp.   _____

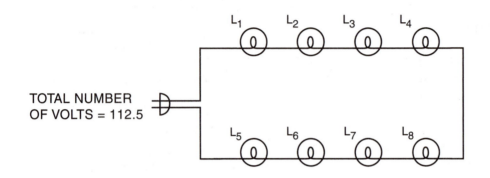

11. If 1,725 feet of Romex cable sell for $286.36, what is the cost of 1,136 feet of this cable?   _____

12. The dry cells shown add up to a total voltage of 27.5 volts. What is the average number of volts for each dry cell?   _____

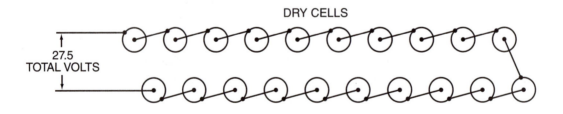

13. Three-conductor, size 0000 cable weighs 3,530 pounds per one thousand feet. Find the weight of 475 feet of this cable.   _____

14. The weight of a certain wire is 145 pounds per one thousand feet. What is the cost of 600 feet of this wire at $1.40 per one hundred pounds?   _____

 # Unit 15 DECIMAL AND COMMON FRACTION EQUIVALENTS

BASIC PRINCIPLES OF DECIMAL AND COMMON FRACTION EQUIVALENTS

Changing Common Fractions to Decimal Fractions

A common fraction is an expression of division. It is possible to change any common fraction into its decimal equivalent by dividing the numerator by the denominator.

Example: Express ⅜ as a decimal.

$$\frac{3}{8} = 3 \div 8 = 0.375$$

Changing Decimal Fractions into Common Fractions

A decimal fraction is a representation of a common fraction that, if converted, has a denominator of 10; 100; 1,000; and so on.

Example: Express 0.00450 as a fraction.

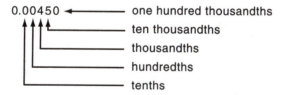

The fraction 0.00450 has a denominator of 100,000. The decimal fraction is written as a common fraction and reduced to lowest terms.

$$\frac{450}{100,000} = \frac{9}{2,000}$$

PRACTICAL PROBLEMS

1. The armature shaft shown, originally 1⅜ inches in diameter, is turned ¹⁄₁₆ inch smaller in diameter due to the irregular surface caused by running in a dry bearing. What is the finished size of the shaft, expressed as a decimal?   _____

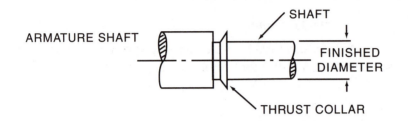

ARMATURE SHAFT

SHAFT

FINISHED DIAMETER

THRUST COLLAR

2. A piece of ¼-inch diameter copper rod is rolled down in a mill to 0.010 inch in thickness for magnet wire sleeving and clips. In decimal form, how much is the copper reduced in size?   _____

3. The total weight of a spool and the magnet wire on the spool is 29.14 pounds. If the spool weighs 3¾ pounds, find the weight of the wire.   _____

4. Express the diameter of a ³⁄₃₂-inch twist drill as a decimal.   _____

5. If a machine screw measures ⅛ inch in diameter, what size hole must a washer have to be 0.006 inch larger than the screw?   _____

6. A meter is to be bolted to a switchboard. The meter studs that will fit into the holes on the switchboard are 0.4365 inch in diameter. Express the hole sizes in decimal form if they are to be ¹⁄₃₂ inch larger in diameter than the studs.   _____

7. The motor bearing shown is to fit a shaft that measures 1.625 inches in diameter. What fractional size reamer should be used to ream a hole in the motor bearing?   _____

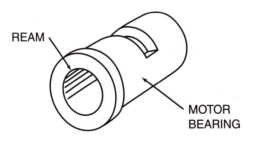

REAM

MOTOR BEARING

8. A bearing with an inside diameter of 1¼ inches is found to be 0.008 inch oversize for the armature shaft. What should the diameter of the bearing be to fit the shaft? Allow 0.002-inch clearance for lubrication.   _____

9. A piece of steel measures 1.375 inches in thickness before grinding, and 1.250 inches after grinding. Express the amount of reduction in fractional form.   _____

10. The carbon brush shown measures ²³⁄₃₂ inch thick and 1¹⁵⁄₃₂ inches wide. It is to have a copper plating on all sides 0.015 inch thick. What will be the finished dimensions in decimal form of the brush with the copper plating?   _____

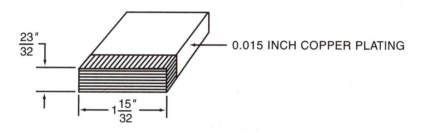

11. A bearing (babbitted) with a 1¾-inch bore is fitted to the new shaft of a motor by reaming. If the reaming process increases the diameter of the bore by 0.003 inch, what is the size of the bore after the work is completed?   _____

12. A set of 2-inch-wide by ⅝-inch-thick carbon brushes fits the brush holders too tightly. It is necessary to take ⁵⁄₁₀₀₀ inch off the width and 0.015 inch off the thickness by sanding. Give in decimal form the width and thickness dimensions of the carbon brushes after sanding.   _____

13. A steel sleeve bearing that has an outside diameter of 2½ inches and an inside diameter of 2⅛ inches must have a babbitt inserted to fit a shaft that has a diameter of 1.609 inches. What is the thickness of the babbitt shown? Express the answer in decimal form.   _____

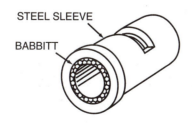

14. A piece of 0.0625-inch-thick copper is used to make a sleeve for a joint. What is the thickness of this material expressed as a fractional part of an inch?   _____

15. The key shown in the figure is ⅞₆ inch wide and ⅝ inch deep. It is found that the key is too large in width (W) and depth (H). The width is machined off 0.005 inch, and 0.003 inch is taken off the depth. What are the dimensions of the key, expressed as decimals, after machining?

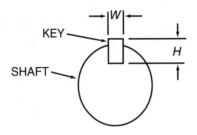

Unit 16 COMBINED OPERATIONS WITH DECIMAL FRACTIONS

This unit provides practical problems involving combined operations of addition, subtraction, multiplication, and division of decimal fractions.

PRACTICAL PROBLEMS

1. What is the total thickness of the following shims taken from a bearing: 0.065 inch, 0.150 inch, 0.130 inch, 0.185 inch, and 0.005 inch? _____

2. What is the total cost to repair an electric outlet if the following materials and labor are required: one shallow box at $0.65, one ⅜-inch hickey at $0.98, one flush toggle switch at $1.17, and one hour of labor at $65? _____

3. What is the total number of amperes in a parallel circuit if the following lamps are connected to the circuit: one 100-watt lamp, 0.834 ampere; one 60-watt lamp, 0.437 ampere; one 40-watt lamp, 0.375 ampere; one 25-watt lamp, 0.225 ampere; one 10-watt lamp, 0.175 ampere; and one 7-watt lamp, 0.125 ampere? _____

4. What is the total thickness of insulation between the copper conductor and the iron core of a motor armature if there is 0.003 inch of enameled insulation, 0.065 inch of cotton insulation, 0.015 inch of slot insulation, and 0.007 inch of varnished cambric insulation? _____

Note: Use this illustration for problems 5–10.

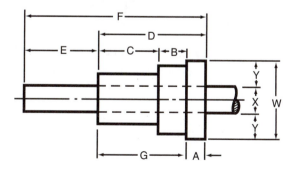

5. If A = 0.305 inch, B = 0.870 inch, and C = 1.0425 inch, find D.   _____

6. If X = 0.9875 inch and Y = 0.675 inch, find W.   _____

7. If E = 3.165 inches and F = 7.025 inches, find D.   _____

8. If B = 0.500 inch and G = 2.0222 inches, find C.   _____

9. If A = 0.305 inch, B = 0.870 inch, and D = 3.860 inches, find G.   _____

10. If A = 0.305 inch, C = 1.0425 inches, and D = 3.860 inches, find B.   _____

11. If the total weight of a spool of number 16 wire is 15.625 pounds and the weight of the spool alone is 7.25 pounds, find the weight of the wire.   _____

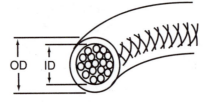

12. The actual inside diameter (ID) of a 3-inch conduit is 3.375 inches, and the actual outside diameter (OD) is 3.9375 inches. What is the wall thickness of this conduit?   _____

13. The field current of a motor is 7.25 amperes at no load. However, when the full load is applied, the current increases to 8.75 amperes. What is the difference in current from no load to full load?   _____

14. What is the total weight of a container of 7 gallons of liquid insulating material if one gallon weighs 9.36 pounds and the container alone weighs 6.87 pounds?   _____

15. If the cost of Romex cable is $108.75 per one hundred feet, determine the total cost for 37 feet.   _____

16. Determine the circumference (C) of a grinding wheel if the diameter (d) is 12 inches. Use the following formula.   _____

$$C = \pi d \quad \text{where} \quad \pi = 3.1416$$

17. An electrician works a total of 40 hours on a renovation job and is paid a total of $980.00. Determine the hourly rate of pay.   _____

18. A certain type of cable weighs 500.88 pounds per one thousand feet. Determine the weight of 53 feet of this cable. Round the answer to the nearer hundredth.   _____

19. The holes drilled in the meter board shown are equally spaced across the board. The length of the board (L) is 56 inches, and the diameter (d) of each hole is 2.75 inches. The distance from each edge of the panel to the nearest hole is the same as the distance (D) between the holes. Find D. Round the answer to the nearer hundredth.   _____

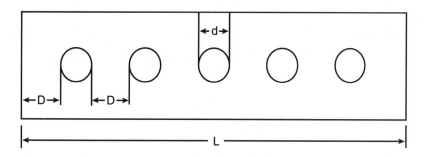

20. Standard 1-inch conduit has an inside diameter of 1.575 inches and an outside diameter of 1.825 inches. Find the thickness of the wall.   _____

Percents, Averages, and Estimates

Unit 17   PERCENT AND PERCENTAGE

BASIC PRINCIPLES OF PERCENT AND PERCENTAGE

Percent means hundredths or *number per 100.* If an electrician has 100 receptacle boxes and uses 45 of them on a job, 45% of the supply is used on that job.

Changing a Decimal Fraction to a Percent

To change a decimal fraction to a percent, move the decimal point two places to the right (multiply by 100) and add a percent sign.

Example: Express 0.40 as a percent.

$$0.40 = 40\%$$

Changing a Common Fraction to a Percent

To change a common fraction to a percent, first change the common fraction to a decimal fraction by dividing the numerator by the denominator. Then move the decimal point two places to the right (multiply by 100) and add a percent sign.

Example: Express ⅝ as a percent.

$$\frac{5}{8} = 0.625 = 62.5\%$$

Changing a Percent to a Decimal Fraction

To change a percent to a decimal fraction, move the decimal point two places to the left (divide by 100) and drop the percent sign.

Example: Express 5% as a decimal fraction.

$$5\% = 0.05$$

Changing a Percent to a Common Fraction

To change a percent to a common fraction, first change the percent to a decimal fraction. Then change the decimal fraction to a common fraction and reduce to the lowest terms.

Example: Express 12.5% as a common fraction.

$$12.5\% = 0.125 = \frac{125}{1000} = \frac{1}{8}$$

Finding What Percent One Number Is of Another

To find what percent one number is of another, first change the numbers to a fraction and then change the fraction to a percent.

Example: 40 is what percent of 50?

$$\frac{40}{50} = 0.80 = 80\%$$

Example: What is 50% of 100? Change 50% into a decimal fraction and multiply.

$$0.50 \times 100 = 50$$

Example: A motor is rated to operate at 480 volts. If the line voltage is 110%, what is the voltage connected to the motor?

$$110\% = 1.10$$
$$480 \times 1.10 = 528 \text{ volts}$$

PRACTICAL PROBLEMS

1. The generator shown ordinarily generates 1,500 volts. Find the percent of voltage increase that it is presently generating.   _____

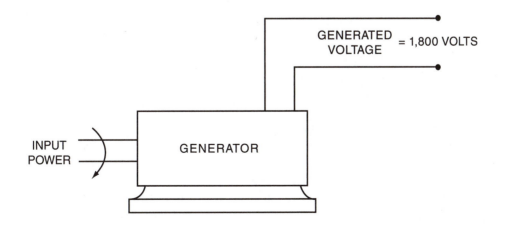

2. A motor rated at 90 horsepower is actually developing 105 horsepower. What is the percent of horsepower overload?

    _____

3. Each worker receives $122.35 per day. The wages are reduced 8%. Find to the nearer cent the amount each worker receives per day after the cut in pay.

    _____

4. In replacing 55 lamp bulbs, an apprentice breaks 6. Find to the nearer hundredth percent the percent broken.

    _____

5. Seven and one-half percent of a group of 640 motors are found to be overloaded. How many are not overloaded?

    _____

6. If $168.00 is the profit on a job, and this represents 8% of the contract price, what is the contract price?

    _____

7. An electrical repairer charges 33% of the cost of a new motor for a rewinding job. If the motor costs $287.00 when new, what is the amount charged for rewinding?

    _____

8. Resistors R_1 and R_2 together use 63% of the total voltage. What is the voltage drop across resistor R_3?

    _____

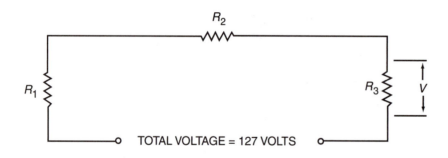

TOTAL VOLTAGE = 127 VOLTS

9. A motor that ordinarily delivers 40 horsepower is delivering only 36 horsepower. Find the percent that is now being delivered as compared to the usual amount.

    _____

10. A 12-volt battery has had a capacity of 30 ampere-hours, but due to aging, has dropped to a capacity of 24 ampere-hours. Find the percent decrease in capacity.

    _____

11. An electrician charges $425.00 for a wiring job. The cost of materials amounts to 62% of the total cost. Find the amount of money that the electrician receives for labor.

    _____

12. Find, to the nearer hundredth horsepower, the amount of input horsepower required for the machine shown if it is to deliver an output of 97 horsepower.   _____

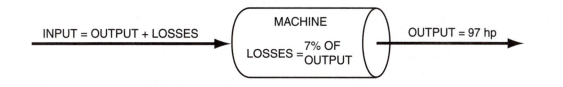

INPUT = OUTPUT + LOSSES

MACHINE

LOSSES = 7% OF OUTPUT

OUTPUT = 97 hp

UNIT 18 INTEREST

BASIC PRINCIPLES OF INTEREST

Interest is a way of using percent. When money is borrowed, the borrowed amount is known as the *principal.* The amount charged for the use of the borrowed money is called the *interest,* and the *rate* of interest is expressed as a percent. Interest is usually computed on a yearly basis (per annum). When the term of the loan has expired, the money repaid is the *sum of the principal and the interest.*

Example: A company borrows $25,000 at an interest rate of 16% for a period of one year. How much money is repaid at the end of one year?

First, change the percent into a decimal fraction.

$$16\% = 0.16$$

Multiply the principal by the decimal fraction to find the interest charge.

$$\$25,000 \times 0.16 = \$4,000$$

Add the interest charge to the principal to find the total amount of money repaid.

$$\$25,000 + \$4,000 = \$29,000$$

If the time of the loan is longer or shorter than one year, divide the annual interest by 12 months and multiply by the number of months. For example, assume the company is unable to repay the loan at the end of 12 months and asks for a 6-month extension. Find the total amount of money repaid at the end of 18 months.

First, determine the interest for 12 months.

$$\$25,000 \times 0.16 = \$4,000$$

Divide the annual interest by 12 months to find the interest charge per month.

$$\$4,000 \div 12 \text{ months} = \$333.33 \text{ per month}$$

Multiply the charge per month by the total number of months to find the total interest.

$$\$333.33 \times 18 = \$6,000$$

Add the total interest and principal to find the amount repaid.

$$\$25,000 + \$6,000 = \$31,000$$

PRACTICAL PROBLEMS

1. A contractor borrows $900 at 12% interest rate per annum. The debt is paid in 1 year and 7 months. Referring to the figure, find the amount paid back at the end of 1 year and 7 months. _____

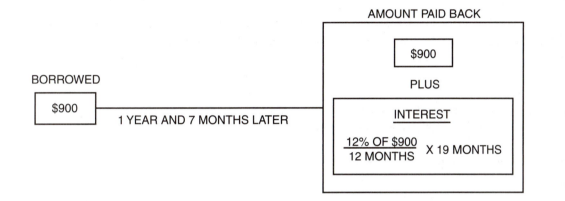

2. A contractor borrows $750 at 12% interest rate per annum and pays the debt 6 months later. What is the interest charged? _____

3. A contractor borrows $2,700 on July 1 at 12% interest rate per annum to purchase electrical supplies. If the loan is repaid 18 months later, how much interest is charged? _____

4. A contractor's note, given 2 years 9 months ago for $856.25 and bearing an 8% interest rate per annum, is paid in full. What is the total amount of interest charged? _____

5. On June 1, five bills for electrical goods are sent out for the following amounts: $1,046.25, $952.40, $164.00, $1,150.00, and $518.00. On December 1 of the same year, payment is received in full, with interest at the rate of 12% per annum. What is the total amount received? _____

6. An electrician borrows $1,250 at 12.5% interest rate per annum to replace direct current motors with alternating current motors. The loan is paid back in 1 year and 2 months. Find the amount of interest. _____

7. A firm purchases heavy cable and 4-inch conduit on credit and agrees to pay 10% interest rate per annum. Purchases are made in October for cable costing $756.80 and in December for conduit costing $1,325.25. If full payment is made by March 1 of the following year, how much is paid? (Charge interest for the full month of purchase.) _____

8. A firm purchases several motors at a cost of $8,350 and gives a note bearing an 8% interest rate per annum. The note is due one year later. What is the total amount to be paid on the note? _____

9. On July 17, a contractor borrows $1,150 at an interest rate of 10% per annum and pays the note when it becomes due 39 months later. What is the amount of interest? _____

10. An electrician purchases a new truck for the amount shown. He trades in an old truck and pays an amount of money in cash toward the total payment. When the loan is paid in full after one year, what will it cost in interest payments? _____

BILL OF SALE	
Cost of New Truck	$24,600.00
Trade-in Allowance	4,500.00
Cash Payment	1,000.00
Balance Due	?
Interest on Balance	12% per Annum
Time to Pay in Full	1 year

11. A contractor has $650 worth of electrical material on the shelf in stock for 3 consecutive years. If the money is borrowed at 12% interest rate per annum, what does it cost to carry this stock? _____

UNIT 19   DISCOUNT

BASIC PRINCIPLES OF DISCOUNT

Discount is another way of using percent. Discount is subtracted from list price to find net price.

Example: A crimp connector tool lists for $18.50. The tool is offered at a 12% discount. What is the net price of the tool?

First, change 12% into a decimal fraction.

$$12\% = 0.12$$

Multiply the list price by the discount.

$$\$18.50 \times 0.12 = \$2.22$$

Subtract the discount from the list price.

$$\begin{array}{r} \$18.50 \\ -\ 2.22 \\ \hline \$16.28 \end{array}$$ (Net price of the tool)

PRACTICAL PROBLEMS

1. An electric pump is listed at $254.25. Find the net cost of the pump at a 20% discount.   _____

2. A contractor purchases quantities of wire, fittings, and switches listed at $2,150 at successive trade discounts of 15%, 10%, and 3%. What is the net cost?   _____

MULTIPLE DISCOUNT

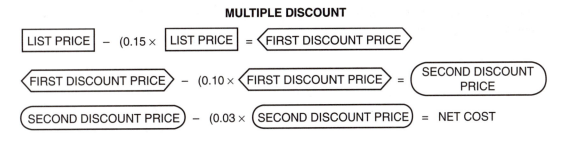

3. Two-inch conduit is listed at $161.56 per hundred feet. An electrician purchases 625 feet of conduit and receives successive trade discounts of 20%, 5%, and 2%. What is the net cost?

    _____

4. Solderless lugs are quoted at $42.50 less 40% for standard packages of 50. What will be the cost if only 22 are ordered at list price less 35%?

    _____

5. Universal 2½-inch conduit caps are listed at $7 each less successive discounts of 20%, 15%, and 2%. Sixteen of these are ordered. What is the total net cost?

    _____

6. A contractor purchases motors listed at $6,800 and receives successive discounts of 20% and 10%. What is the net cost?

    _____

7. An electrician purchases 1,250 feet of two-inch flexible steel conduit at a list price of $109 per hundred feet. With successive discounts of 20% and 10%, what is the net cost?

    _____

8. Three electrical distributors offer the same grade of materials at the same list price. Distributor A offers discounts of 25% and 15%; distributor B offers discounts of 20% and 20%; and distributor C offers discounts of 15%, 15%, and 10%. Materials listed at $350 are to be purchased.

    _____

   a. Find the net cost when the materials are purchased from distributor A.   a. _____

   b. Find the net cost when the materials are purchased from distributor B.   b. _____

   c. Find the net cost when the materials are purchased from distributor C.   c. _____

9. Find the total net cost of the shipment of materials shown.

    _____

ITEM	QUANTITY	LIST	DISCOUNT
Floor Box	10	$4.00 each	20%
½-inch Conduit	250	$2.26 each	25%
Single-Pole Key Socket Body	250	$4.14 each	40%
Single-Pole Flush Switch	35	$1.27 each	40%

10. What is the net price of a 2-pole, 400-ampere, 230-volt enclosed entrance switch if it has a list price of $287 with successive discounts of 48% and 2%? _____

11. Seventy-two 30-ampere, 600-volt enclosed refillable fuse cases are purchased at a list price of $27.50 each with a discount of 28%. What is the total cost? _____

12. An electrician purchases thirty 125-volt, 30-ampere, double-pole, double-branch cutouts listed at $6.40 per box of 5, less 25%, and eight surface panels listed at $4.75 each, less 35%. Three percent is saved by paying the bill in 15 days. What is the cost if paid within 15 days? _____

# UNIT 20   AVERAGES AND ESTIMATES

BASIC PRINCIPLES OF AVERAGES AND ESTIMATES

Averages

Average value is found by adding individual values of some number of units and dividing the sum by the total number of units.

Example: A company truck has the following gasoline expenses:

First week	$68.50
Second week	$87.34
Third week	$103.09
Fourth week	$77.16

What is the average weekly gas bill for this truck?

Find the sum of the total gasoline bills for the month and then divide the total by the number of weeks.

$$
\begin{array}{r}
68.50 \\
87.34 \\
103.09 \\
+\ \ 77.16 \\
\hline
336.09
\end{array}
$$

$336.09 \div 4 = \$84.02$ (average)

Estimates

Estimates are used for approximation and are not intended to be exact. Being able to estimate the cost of materials for a job or the amount of time needed to perform a certain task requires time and experience.

Example: An electrician estimates the materials for a certain job will cost $120. When the job is finished, it is found that the actual cost of materials is $112.16. The actual amount is within what percent of the estimate?

$$\$112.16 \div \$120 = 93.5\% \quad 100\% - 93.5\% = 6.5\%$$

The actual amount is within 6.5% of the estimated amount.

PRACTICAL PROBLEMS

1. The amounts of power used in an electrical maintenance shop are as follows: April, 41.2 kilowatt-hours; May, 59.25 kilowatt-hours; June, 53.63 kilowatt-hours; July, 62.4 kilowatt-hours; August, 63.75 kilowatt-hours; September, 30.35 kilowatt-hours. What is the average monthly power usage? Express the answer to the nearer hundredth.   _____

2. A 208-volt, 4-wire, 3-phase wye power system feeds a school building. Daily voltage readings taken for one week are 205.25 volts, 203.75 volts, 208 volts, 204.35 volts, 206.7 volts, 207 volts, and 208.55 volts. What is the average daily voltage of the system? Round the answer to the nearer hundredth.   _____

3. An electrician earns $25.30 per hour. During one week the electrician works these hours: Monday, 8 hours; Tuesday, 7 hours; Wednesday, 5½ hours; Thursday, 10 hours; and Friday, 4½ hours. What is the average daily earning?   _____

4. An electrician estimates three bundles (approximately 300 feet) of ½-inch conduit are needed to wire a house. Each bundle costs $62.00 or $0.62 per foot. The amounts used are as follows: living room, 42 feet; dining room, 38 feet; kitchen, 115 feet; bedroom A, 24 feet; bedroom B, 39 feet; and bathroom, 12 feet.   _____

    a. What is the average cost per room?   a. _____
    b. What is the electrician's estimated cost per room?   b. _____
    c. The electrician's estimate of the cost is how much over or under the cost   c. _____
        of the job?

5. Five light fixtures for a house cost these amounts: type A, $15.83; type B, $21.75; type C, $19.69; type D, $24.30; and type E, $9.49.   _____

    a. What is the average cost per fixture?   a. _____
    b. The electrician estimates a cost of $19.00 per fixture. How much over or   b. _____
        under is the estimate for the five fixtures?

6. Two-inch angle iron weighing approximately 2.97 pounds per foot is used to build panel-mounting racks. These amounts of iron are used: panel A, 9 feet; panel B, 10.75 feet; panel C, 8.6 feet; and panel D, 7.7 feet. What is the average weight per panel? Round the answer to the nearer hundredth.   _____

7. An electrician estimates 2,500 feet of number 12 NM cable is needed to wire a house. Each coil of cable holds 250 feet. The amounts used in different rooms are as follows: 335.4 feet, 293.7 feet, 1,205.1 feet, and 337.5 feet.

    a. How many coils of wire are used?   a. _____

    b. How many feet over or under is the estimate?   b. _____

8. A shipment of 15 heat pumps arrives for storage at an electrical supply house. The pumps cannot be stacked, and each has a base measurement of 3.2 feet by 3.8 feet. A storage area 14 feet long and 12 feet wide is available. Is there sufficient space to store the shipment?   _____

9. The temperature of a walk-in cold provisions cabinet is sampled twice daily for a 3-day period. The Fahrenheit temperature readings are as follows: 37.0°, 33.0°, 35.5°, 34.0°, 38.0°, and 36.6°. What is the average temperature? Express the answer to the nearer tenth.   _____

10. It is required that a cold storage temperature be maintained at an average of 3.055 degrees Celsius plus or minus 35%.

    a. Find the lowest temperature reading that satisfies this requirement.   a. _____
        Express the answer to the nearer thousandth.

    b. Find the highest temperature reading that satisfies this requirement. b. _____
        Express the answer to the nearer thousandth.

UNIT 21   COMBINED PROBLEMS ON PERCENTS, AVERAGES, AND ESTIMATES

This unit provides practical problems involving combined problems on percents, averages, and estimates.

PRACTICAL PROBLEMS

1. A motor rated at 30 horsepower is actually developing 35 horsepower. What is the percent of horsepower overload? Round to the nearer tenth.   _____

2. Workers receiving $124.60 per day have their wages increased by 8%. How much is received per day after the increase?   _____

3. An engine has an input of 90 horsepower and uses 8% of its input power to overcome friction and other losses. Find the available horsepower at the output if the output = input − losses.   _____

4. When mixing a quantity of electrolyte for a storage battery, the electrician uses 2 parts of acid and 3 parts of water. What percent is acid?   _____

5. A contractor borrows $1,000 at 11.8% interest rate per annum to buy material for a job. The debt is paid 18 months later. What amount is paid in interest?   _____

6. An electrician borrows $2,750 at 9.7% interest rate per annum to purchase electrical supplies. If the loan is repaid in 15 months, how much is the interest?   _____

7. A contractor borrows $3,500 at an interest rate of 6% per annum and pays the note when it becomes due 6 months later. What is the amount of interest paid for the use of the money?   _____

8. A contractor purchases a truck for $16,800 and is allowed $2,000 for trading in an old truck. A cash deposit of $200 is made. How much interest is paid for one year at a rate of 12% per annum?   _____

9. Find the total cost of the following shipment of materials: 23 floor boxes listed at $4.00 each, less 15%; 150 half-inch conduits listed at $0.60 each, less 20%; 125 convenience outlets listed at $0.79 each, less 10%; 83 single-pole flush switches listed at $0.45 each, less 30%.   _____

10. What is the net price of a 2-pole, 100-ampere, 230-volt entrance switch if the list price is $137 with successive discounts of 35% and 3%?   _____

11. An electrician is paid $1,420 in January; $1,560 in February; $1,672 in March; $1,878 in April; $1,925 in May; and $2,016 in June. What is the average monthly pay?   _____

12. An electrical supply uses the following wattages: 4,212 watts, 4,226 watts, 4,296 watts, 3,427 watts, 914 watts, 1,428 watts, and 4,293 watts. What is the average wattage used? Round the answer to the nearer hundredth.   _____

13. An electrician estimates using 500 staples on a six-room house. The electrician uses 70 staples in the first room, 80 staples in the second room, 50 staples in the third room, 75 staples in the fourth room, 95 staples in the fifth room, and 100 staples in the sixth room. Find the percent of accuracy of the estimate.   _____

14. Find the average of the following measurements: 17 inches, 16 inches, 18 inches, 21 inches, 29 inches, and 24 inches. Round the answer to the nearer hundredth.

Powers and Roots

Unit 22   POWERS

BASIC PRINCIPLES OF POWERS

The *power* of a number is also known as its *exponent*. If a number is *squared,* for example, it is raised to the second power, or has an exponent of 2. The exponent is the small number written above and to the right of the number as shown. This means that the number is to be multiplied by itself.

$$5^2 = (5 \times 5) = 25$$

Notice that 5 is not multiplied by 2, but is multiplied by itself.

If a number is *cubed,* it means that the number is raised to the third power, or has an exponent of 3.

$$8^3 = (8 \times 8 \times 8) = 512$$

A number can be raised to any power by using it as a factor that number of times.

Example: Raise 6 to the fifth power.

$$6^5 = (6 \times 6 \times 6 \times 6 \times 6) = 7,776$$

PRACTICAL PROBLEMS

Raise the following expressions to the power indicated.

1. 7^2 _____

2. 8^2 _____

3. 9^2 _____

4. 10^2 _____

5. 11^2 _____

6. 7^3 _____

7. 8^3 _____

8. 9^4 _____

9. 10^4 _____

10. 11^5 _____

11. 15^2 _____

12. 21^2 _____

13. 25^3 _____

15. 30^4 _____

17. 43^5 _____

14. 28^3 _____

16. 35^5 _____

18. 50^6 _____

Note: Refer to the following information for problems 19–21.

For a round conductor, the square of the diameter *(d)* equals the cross-sectional area *(CM)*. That is,

$$d^2 = CM$$

where *d* is measured in mils (1 mil = 0.001 inch) and *CM* is measured in circular mils.

19.   Find the cross-sectional area in circular mils of the wire shown.   _____

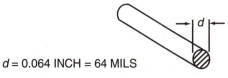

$d = 0.064$ INCH $= 64$ MILS

20.   Find the cross-sectional area in circular mils of a wire that is 104 mils in diameter.   _____

21.   What is the cross-sectional area in circular mils of a conductor that has a diameter of 32 mils?   _____

Note: The symbols used in problems 22–25 are as follows:

P = Power in watts   *I* = Current in amperes
E = Voltage in volts   *R* = Resistance in ohms

22.   To find the number of watts used in any circuit, the amperes squared are multiplied by ohms. Find the number of watts for the device shown.   _____

$$P = I^2R = (\text{amperes})^2\ (\text{ohms}) = \text{watts}$$

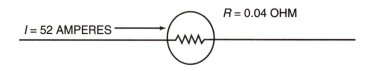

$R = 0.04$ OHM

$I = 52$ AMPERES

23. Find the wattage for the circuit shown.   _____

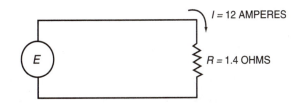

I = 12 AMPERES

R = 1.4 OHMS

E

24. The number of watts used in a circuit is equal to the volts squared divided by the ohms. Find the wattage used in the circuit shown. Round the answer to the nearer tenth.   _____

$$P = \frac{E^2}{R} = \frac{(\text{volts})^2}{\text{ohms}} = \text{watts}$$

E = 100 VOLTS

R = 1.3 OHMS

25. If the voltage in Problem 24 is changed to 220 volts and the resistance is changed to 3.6 ohms, how many watts are used? Round the answer to the nearer tenth.   _____

26. A cable is checked for a power installation and is found to be too small. The cable has 37 strands of copper wire, each having a diameter of 82.2 mils. Find the size of the cable in circular mils.   _____

27. If a flexible copper cable has 133 strands, each 5.63 mils in diameter, what is the size of the cable in circular mils?   _____

 # UNIT 23 ROOTS

BASIC PRINCIPLES OF ROOTS

The *root* of a number is the opposite or inverse of its power. The *square root* of a number is determined by finding which number used as a factor twice will equal the given number. The square root of 36 is 6 because 6 squared equals 36.

$$(6 \times 6) = 36$$

The process is similar for other roots also. The *cube root* of 27 is 3 because 3 cubed equals 27.

$$(3 \times 3 \times 3) = 27$$

When the root of a number is to be found, the number is placed under a *radical sign* ($\sqrt{\ }$). If the square root of a number is to be found, the problem is written like this:

$$\sqrt{36} = 6$$

If some root other than a square root is to be found, a small number is placed outside the radical sign. This number indicates what root is to be found.

$$\sqrt[3]{27} = 3$$

FINDING ROOTS WITH A CALCULATOR

Most scientific calculators can be used to find roots of numbers. Almost all calculators contain a square root key ($\sqrt{x}$). To find the square root of a number, simply input the number and press the $\sqrt{x}$ key.

Example: Find the square root of 441.

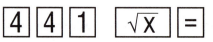

      The answer is 21.

It is sometimes necessary to find other roots of numbers. Assume, for example, that you need to find the fourth root of 50,625 ($\sqrt[4]{50,625}$). How you do this will depend on the type of calculator you use. Although most scientific calculators have the ability to find any root of any number, not all do it the same way. Some calculators contain an *nth* root key:

To find the fourth root of 50,625 using a calculator with an nth root key, input 50625, press the nth root key, press 4, and then press =.

Example:

The answer is 15.

If the calculator does not contain an nth root key, it will almost certainly contain an nth power key:

There are two methods of using the nth power key to find an nth root. If the calculator contains an invert key (INV), it can be used to change the nth power function into an nth root function. To find the fourth root of 50,625 using a calculator of this type, follow the procedure shown:

Example:

For calculators that do not contain the invert key, the nth power key can be used to find an nth root by using the reciprocal key ($\frac{1}{x}$). Any number raised to the reciprocal of any power will be the same as finding the root. Raising 50,625 to the reciprocal of the one-fourth power ($\frac{1}{4}$ = 0.25) is the same as finding the fourth root of 50,625. That is,

$$\sqrt[4]{50,625} = 50,625^{\frac{1}{4}} = 50,625^{0.25}$$

Example:

The answer is 225.

PRACTICAL PROBLEMS

Solve the following expressions, and express the answer to the nearer hundredth.

1. $\sqrt{81}$ _____

2. $\sqrt{100}$ _____

3. $\sqrt{169}$ _____

4. $\sqrt{361}$ _____

5. $\sqrt{529}$ _____

6. $\sqrt{743}$ _____

7. $\sqrt{892}$ _____

8. $\sqrt{1,235}$ _____

9. $\sqrt{1,692}$ _____

10. $\sqrt{2,000}$ _____

Note: P = Power in watts I = Current in amperes

E = Voltage in volts R = Resistance in ohms

11.   The circuit shown uses 2,048 watts. How many amperes are in the circuit?   _____

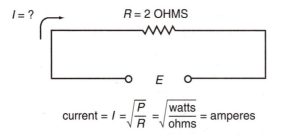

$$\text{current} = I = \sqrt{\frac{P}{R}} = \sqrt{\frac{\text{watts}}{\text{ohms}}} = \text{amperes}$$

12.   An electrical cable has a cross-sectional area *(CM)* of 1,000,000 circular mils and contains 19 strands. Find the diameter *(d)* of each strand, using   _____

$$d = \sqrt{CM \text{ per strand}}$$

13.   An electric heating device uses power in the amount of 605 watts. The resistance of the device is 20 ohms. This heater operates on what voltage?   _____

$$E = \sqrt{P \times R}$$

Note: Use the circuit and the associated formulas shown for problems 14–22.

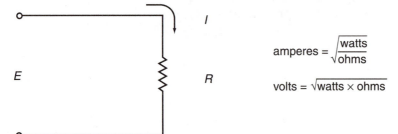

$$\text{amperes} = \sqrt{\frac{\text{watts}}{\text{ohms}}}$$

$$\text{volts} = \sqrt{\text{watts} \times \text{ohms}}$$

14.   If the resistance has a value of 550 ohms and uses 22 watts of power, find the number of amperes of current that the resistance takes.   _____

15.   A circuit using 880 watts of power with 220 ohms of resistance requires what value of current in amperes?   _____

16.   How many volts are required if the power is 25 watts and the resistance is 64 ohms?   _____

17. What current in amperes is present in the circuit if the resistance is 0.88 ohm and the power is 13.75 watts?

_____

18. When the power is 220 watts and the resistance is 100 ohms, what voltage is in the circuit?

_____

19. When the resistance is 495 ohms and the power is 22 watts, what is the value of current in amperes?

_____

20. When the resistance is 73.3 ohms and the power is 165 watts, what is the value of the current in amperes?

_____

21. The resistance uses 287 watts and has a value of 46 ohms. How many amperes exist in the circuit?

_____

22. Find the number of volts in the circuit if the resistance has a value of 22 ohms and uses 550 watts when the current value is 5 amperes.

_____

23. The cross-sectional area in circular mils of a number 10 wire is 10,380. The diameter in mils of a wire is found by taking the square root of the cross-sectional area. Find the diameter of the number 10 wire.

_____

24. If a 250,000-circular-mil cable is composed of 37 strands, what is the approximate diameter of each strand?

_____

25. To find the line voltage in the 2-phase, 3-wire circuit shown, the constant $\sqrt{2}$ is used. The line voltage is equal to $\sqrt{2}$ multiplied by voltage *(E)*. Find the value of the line voltage for this circuit.

_____

$$E_{LINE} = \sqrt{2} \times E$$

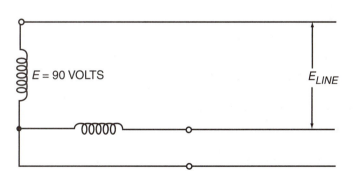

26. A 5,000,000-circular-mil cable is composed of 37 strands of copper wire of equal size. What is the approximate diameter of one strand?   _____

27. The square root of 3 is used to make calculations for 3-phase, 3-wire circuits like the one shown. The line current for this type of circuit is equal to $\sqrt{3}$ multiplied by current (I). Find the line current for this circuit.   _____

$$I_{LINE} = \sqrt{3} \times I$$

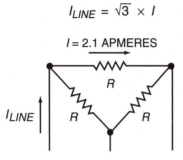

28. If the cross-sectional area in circular mils of a wire is 4.107, what is the diameter in mils?   _____

29. A 750,000-circular-mil cable is composed of 61 strands of copper wire of equal size. What is the approximate diameter of one of the strands?   _____

UNIT 24   COMBINED OPERATIONS WITH POWERS AND ROOTS

This unit provides practical problems involving combined operations with powers and roots.

PRACTICAL PROBLEMS

1. $\sqrt{121}$ _____   6. $\sqrt{1,156}$ _____

2. $\sqrt{196}$ _____   7. $\sqrt{1,521}$ _____

3. $\sqrt{289}$ _____   8. $\sqrt{1,936}$ _____

4. $\sqrt{441}$ _____   9. $\sqrt{3,136}$ _____

5. $\sqrt{729}$ _____   10. $\sqrt{4,356}$ _____

11. The diameter of number 10 copper wire is 101.9 mils. Find the cross-sectional area of the wire in circular mils. _____

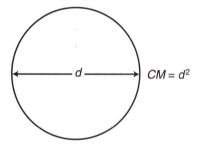

$CM = d^2$

12. A cable is checked for a power job and is found to be too small. The cable contains 62 strands of copper wire that each have a diameter of 37.2 mils. What is the size of the cable in circular mils? _____

13. The cross-sectional area in circular mils of a number 12 wire is 6,530. Find the diameter. Round the answer to the nearer mil. _____

14. A 2,583-circular-mil cable is composed of 25 strands of copper wire of equal size. What is the approximate diameter of one strand? Round the answer to the nearer tenth. _____

15. A circuit has a power rating of 500 watts and a resistance of 65 ohms. What is the current flow in the circuit? Round the answer to the nearer hundredth of an ampere.

_____

$$I = \sqrt{\frac{P}{R}} \qquad \text{amperes} = \sqrt{\frac{\text{watts}}{\text{ohms}}}$$

16. The resistance of a coil of wire that uses 800 watts is 62 ohms. What is the voltage of the circuit? Round the answer to the nearer hundredth.

_____

$$E = \sqrt{P \times R} \qquad \text{volts} = \sqrt{\text{watts} \times \text{ohms}}$$

17. If a circuit uses 700 watts and has a resistance of 50 ohms, what is the amperage of the circuit? Round the answer to the nearer hundredth.

_____

$$I = \sqrt{\frac{P}{R}}$$

18. 480 volts is connected to an 18 ohm resistive-heating element. How much power is being produced by the heating element?

_____

19. A 5 kilowatt heating element is powered by a 240 volt line. What is the resistance of the element?

_____

20. A 24 ohm resistor has a current of 0.65 amperes flowing through it. How much power is being produced by the resistor?

_____

UNIT 25   METRIC MEASURE AND SCIENTIFIC NOTATION

Metric measurement is used throughout the scientific community. Metric units of measure progress in steps of ten. Common metric units are deka, hector, kilo, deci, centi, and milli. Deka means 10, hecto means 100, and kilo means 1,000. Deci means $\frac{1}{10}$, centi means $\frac{1}{100}$, and milli means $\frac{1}{1,000}$. Standard units of metric measure are shown in the first chart below. A dekameter, for example, would be ten meters. A decimeter would be $\frac{1}{10}$ of a meter.

Although metric units progress in steps of ten, electrical measurements progress in steps of 1,000. These units of measure are generally referred to as *engineering units.* Common engineering units above the base unit are kilo, mega, giga, and terra. Engineering units below the base unit are milli, micro, nano, and pico, as shown in the second chart below. For example, one thousand ohms equals one kilo ohm. One thousand kilo ohms (kΩ) equals one megohm (MΩ).

Standard Units of Metric Measure			
Kilo	**1000**	**Deci**	$\frac{1}{10}$ **or 0.1**
Hecto	**100**	**Centi**	$\frac{1}{100}$ **or 0.01**
Deka	**10**	**Milli**	$\frac{1}{1,000}$ **or 0.0001**
	Base unit 1		

Standard Engineering Units			
ENGINEERING UNIT	**SYMBOL**	**MULTIPLY BY:**	
Tera	T	1,000,000,000,000	$\times 10^{12}$
Giga	G	1,000,000,000	$\times 10^{9}$
Meg	M	1,000,000	$\times 10^{6}$
Kilo	K	1,000	$\times 10^{3}$
Base unit		1	
Milli	m	.001	$\times 10^{-3}$
Micro	μ	.000001	$\times 10^{-6}$
Nano	n	.000000001	$\times 10^{-9}$
Pico	p	.000000000001	$\times 10^{-12}$

SCIENTIFIC NOTATION

Scientific notation is used in almost all scientific calculations. It was first used for making calculations with a slide rule. A slide rule is a tool that can perform mathematical operations such as multiplication and division. It can be used to find square roots and logs of numbers, as well as sines, cosines, and tangents of angles. When finding a number on a slide rule, only the actual digits are used, not decimal points or zeros (unless they come between two other digits, such as 102). To a slide rule, the numbers 0.000012, 0.0012, 0.12, 1.2, 12, 120, 12,000, or 12,000,000 are all the same number: 12.

Since the slide rule recognizes only the basic digits of any number, imagine the problem of determining where to place a decimal point in an answer. As long as only simple calculations are done, there is no problem in determining where the decimal point should be placed.

Example: Multiply 12×20

It is obvious where the decimal point should be placed in this problem.

$$240.00$$

Example: Now assume that the following numbers are to be multiplied together.

$$0.000041 \times 380,000 \times 0.19 \times 720 \times 0.0032$$

In this problem, it is not obvious where the decimal point should be placed in the answer. Scientific notation can be used to simplify the numbers so that an estimated answer can be obtained, by dividing or multiplying numbers by the power of 10 to obtain a simple whole number. **Any number can be multiplied by 10 by moving the decimal point one place to the right. Any number can be divided by 10 by moving the decimal point one place to the left.** For example, the decimal fraction 0.000041 can be changed to the whole number 4.1 by multiplying it by 10 five times. Therefore, if the number 4.1 is divided by 10 five times, it will be the same as the original number, 0.000041. In this problem, the number 0.000041 will be change to 4.1 by multiplying it by 10 five times. The new number is 4.1 times 10 to the negative fifth:

$$4.1 \times 10^{-5}$$

Since superscripts and subscripts are often hard to produce in printed material, it is a common practice to use the letter E to indicate the exponent in scientific notation. The above notation can also be written: 4.1E–05.

The number 380,000 can be reduced to 3.8 by dividing it by 10 five times. The number 3.8 must therefore be notated to indicate that the original number is actually 3.8 multiplied by 10 five times.

$$3.8 \times 10^{5} \text{ or } 3.8E05$$

The other numbers in the problem can also be changed to simpler whole numbers using scientific notation: 0.19 becomes 1.9E–01, 720 becomes 7.2E02, and 0.0032 becomes 3.2E–03. Now that the numbers have been simplified using scientific notation, an estimate can be obtained by multiplying these simple numbers together and adding the exponents: 4.1 is about 4; 3.8 is about 4; 1.9 is about 2; 7.2 is about 7; and 3.2 is about 3.

Therefore, $4 \times 4 \times 2 \times 7 \times 3 = 672$.

When the exponents are added:

(E–05) + (E05) + (E–01) + (E02) + (E–03) = E–02

The estimated answer becomes 672E–02. When the calculation is completed, the actual answer becomes:

682.03E–02 or 6.8203

USING SCIENTIFIC NOTATION WITH CALCULATORS

In the early 1970s, scientific calculators, often referred to as "slide rule calculators," became commonplace. Most of these calculators have the ability to display from eight to ten digits, depending on the manufacturer. Scientific calculations, however, often involve numbers that contain more than eight or ten digits. To overcome the limitation of an eight- or ten-digit display, slide rule calculators depend on scientific notation. When a number becomes too large for the calculator to display, scientific notation is used automatically. Imagine that it became necessary to display the distance (in kilometers) that an object would travel in one year at the speed of light (approx. 9,460,800,000,000 km). This number contains thirteen digits. The calculator would display this number as shown.

9.4608 12

The number 12 shown to the right of 9.4608 is the scientific notation exponent. This number could be written 9.4608E12, indicating that the decimal point should be moved to the right twelve places.

If a minus sign should appear ahead of the scientific notation exponent, it indicates that the decimal point should be moved to the left. The number on the display shown below contains a negative scientific notation exponent.

7.5698 –06

This number could be rewritten as shown.

0.0000075698

ENTERING NUMBERS IN SCIENTIFIC NOTATION

Scientific calculators also have the ability to enter numbers in scientific notation. To do this, the exponent key must be used. There are two ways that most manufacturers mark the exponent key: EXP and EE.

Assume that the number 549E08 was to be entered. The following key strokes would be used:

$\boxed{5}$ $\boxed{4}$ $\boxed{9}$ $\boxed{\text{EE}}$ $\boxed{8}$

The calculator would display the following:

549 08

If a number with a negative exponent is to be entered, the "change sign" (+/–) key should be used. Assume that the number 1.276E-4 is to entered. The following key strokes would be used:

The display would show the following:

1.276   –04

SETTING THE DISPLAY

Some calculators permit the answer to be displayed in any of three different ways. One of these ways is with "floating decimals" (Flo). When the calculator is set for this mode of operation, the answers will be displayed with the decimal point appearing in the normal position. The only exception would be if the number to be displayed is too large. In this case, the calculator will automatically display the number in scientific notation.

In the Scientific mode (Sci), the calculator will display all entries and answers in scientific notation.

When set in the Engineering mode (Eng), the calculator will display all entries and answers in scientific notation, but only in steps of three or in multiples of one thousand. When displayed in steps of three, the notation corresponds to standard engineering notation units such as kilo, mega, giga, milli, micro, and so on. For example, assume a calculator is set in the Scientific mode and displays the number shown.

5.69836   05

Now assume that the calculator is reset to the Engineering mode. The number would now be displayed as shown:

569.836   03

The number could now be read as "569.836 kilo," because kilo means one thousand or 1E03.

PRACTICAL PROBLEMS

1. A Canadian football field is 100 meters in length. How many hectometers is this? _____

2. A radio signal is measured at 150 microvolts (μV). Express this as volts. _____

3. In order to correct the power factor of a motor, 0.00035 farads or capacitance must be connected in parallel with the motor. Express this in microfarads. _____

4. A scientific calculator is set to operate in the Floating Decimal mode (Flo). The calculator display shows 945635.045. If the calculator is changed to operate in the Scientific mode (Sci), what would the display show?   _____

5. Assume that the calculator in question 4 is reset to operate in the Engineering mode (Eng). What would the display show?   _____

6. The marked value of a capacitor is 470 pF. Write this value in farads.   _____

7. The average microwave oven operates at a frequency of 2.45 GHz. Express this value in hertz.   _____

8. A calculator is set in the engineering mode. The display shows: 56.345 −03. Express this value in the proper engineering unit.   _____

9. A power plant has a generating capacity of 3.5 gigawatts. Express this value as kilowatts.   _____

10. An industrial plant is supplied with high voltage power from the power company. The power lines are rated 13.8 kV. Express this value as volts.   _____

Measure

 ## Unit 26 LENGTH MEASURE

BASIC PRINCIPLES OF LENGTH MEASURE

The English System of Measure

The English system of measure was originally based on the measurements of the king's body and other natural things. The length of the king's foot, for example, was the standard length of a foot. The middle joint of the king's index finger was the standard length of one inch. Another weight measurement still used today is the grain. A grain was the weight of a grain of wheat.

These measurements are impossible to duplicate because no two things that occur in nature are exactly the same. No two grains of wheat weigh exactly the same, for instance. Today, there are standards of measurement that are kept by the International Bureau of Weights and Standards. The length of a foot, for example, is always the same.

The Metric System

The standard length in the metric system is the *meter*. The length of the meter was originally based on a measurement of the earth. Today, its standard is the length of a certain number of wave lengths in a ray emitted by the element krypton. The metric system is also based on a standard value of 10. This is an advantage over the English system. Lengths of English and metric measure are shown in the following charts.

ENGLISH LENGTH MEASURE		
12 inches (in)	=	1 foot (ft)
3 feet (ft)	=	1 yard (yd)
1,760 yards (yd)	=	1 mile (mi)
5,280 feet (ft)	=	1 mile (mi)

METRIC LENGTH MEASURE		
10 millimeters (mm)	=	1 centimeter (cm)
10 centimeters (cm)	=	1 decimeter (dm)
10 decimeters (dm)	=	1 meter (m)
10 meters (m)	=	1 dekameter (dam)
10 dekameters (dam)	=	1 hectometer (hm)

ENGLISH–METRIC EQUIVALENTS LENGTH MEASURE		
1 inch (in)	≈	25.4 millimeters (mm)
1 inch (in)	≈	2.54 centimeters (cm)
1 foot (ft)	≈	0.3048 meter (m)
1 yard (yd)	≈	0.9144 meter (m)
1 mile (mi)	≈	1.609 kilometers (km)
1 millimeter (mm)	≈	0.03937 inch (in)
1 centimeter (cm)	≈	0.39370 inch (in)
1 meter (m)	≈	3.28084 feet (ft)
1 meter (m)	≈	1.09361 yards (yd)
1 kilometer (km)	≈	0.62137 mile (mi)

Note: ≈ indicates "approximately equals"

Example: A length of wire measures 84 feet. How many meters is this?

$$84 \times 0.3048 = 25.603 \text{ meters}$$

Example: A Canadian football field is 100 meters in length. How long is this in yards?

$$100 \times 1.09361 = 109.361 \text{ yards}$$

PRACTICAL PROBLEMS

1. Find the number of millimeters in 1.2 inches. _____

2. Find the number of centimeters in 54 millimeters. _____

3. Find the number of millimeters in 3.7 centimeters. _____

4. Find the number of meters in 4,376 millimeters. _____

5. Find the number of meters in 43 centimeters. _____

6. Express ¾ inch to the nearer hundredth millimeter. _____

7. Express 6 inches to the nearer hundredth centimeter. _____

8. Express 9 feet to the nearer thousandth meter. _____

9. Express 55 miles to the nearer thousandth kilometer. _____

10. Express 100 yards to the nearer hundredth meter. _____

11. A wiring job requires 73 meters of 3.8-centimeter EMT. There are 9 pieces of the required EMT on hand. Each piece on hand measures 10 feet. How many additional 10-foot lengths need to be ordered for this job? _____

12. The outside diameter of an electrical conduit is 6.3 centimeters. The thickness of the conduit is 0.4 centimeter. Find the inside diameter of the conduit. _____

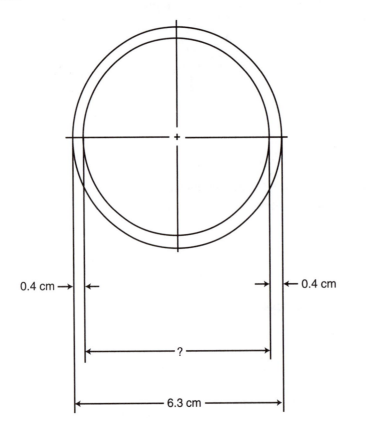

0.4 cm → ← 0.4 cm

?

6.3 cm

Note: Use this illustration for problems 13 and 14.

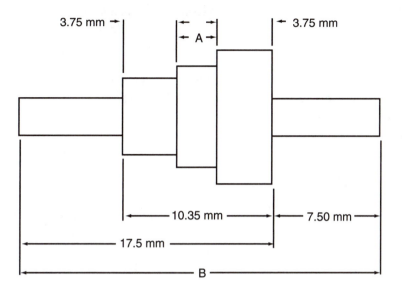

13. Find dimension A in millimeters.   _____

14. Find dimension B in centimeters.   _____

15. A concrete pad 5 meters by 7 meters is poured adjacent to a wall to support a light dimmer. The dimmer is 91 centimeters wide, 457 centimeters long, and 366 centimeters high. The dimmer is placed with a 1-meter clearance from the wall. What is the distance A in meters across the width of the pad from the base of the dimmer to the edge of the pad?   _____

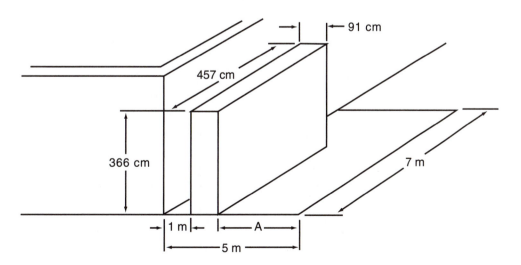

16. To run a 3-wire cable, a hole is drilled through a plate, the subflooring, and the finished flooring. The plate is 1.6 centimeters, and the subflooring is 3.4 centimeters. The finished flooring is 8 millimeters. Find the depth of the hole in centimeters.

_____

17. A hole is cut in a Sheetrock wall for installation of a junction box that is 7.62 centimeters wide and 12.7 centimeters long. A clearance of 0.3 centimeter is required on all sides of the box.

   a. What is the width of the hole?   a. _____

   b. What is the length of the hole?   b. _____

18. Find the distance A in the eccentric cam shown.   _____

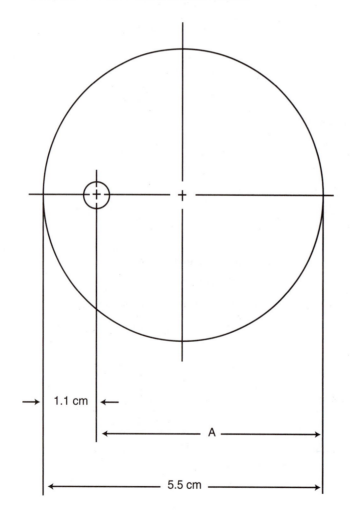

1.1 cm

A

5.5 cm

 **_UNIT 27   AREA MEASURE_**

BASIC PRINCIPLES OF AREA MEASURE

The charts that follow show area measurements for both the English and metric systems. The measurement of area is a two-dimensional quantity. The measurement of area is found by multiplying length and width. A good example of area is the size of the image projected by a slide projector. The length and width of the image can be measured.

ENGLISH AREA MEASURE
144 square inches (sq in) = 1 square foot (sq ft)
9 square feet (sq ft)   = 1 square yard (sq yd)

METRIC AREA MEASURE		
100 square millimeters (mm^2)	=	1 square centimeter (cm^2)
100 square centimeters (cm^2)	=	1 square decimeter (dm^2)
100 square decimeters (dm^2)	=	1 square meter (m^2)
100 square meters (m^2)	=	1 square dekameter (dam^2)
100 square dekameters (dam^2)	=	1 square hectometer (hm^2)
100 square hectometers (hm^2)	=	1 square kilometer (km^2)

ENGLISH–METRIC EQUIVALENTS LENGTH MEASURE		
1 square inch (sq in)	$\approx$	645.16 square millimeters (mm^2)
1 square inch (sq in)	$\approx$	6.4516 square centimeters (cm^2)
1 square foot (sq ft)	$\approx$	0.092903 square meter (m^2)
1 square yard (sq yd)	$\approx$	0.836127 square meter (m^2)
1 square millimeter (mm^2)	$\approx$	0.001550 square inch (sq in)
1 square centimeter (cm^2)	$\approx$	0.15500 square inch (sq in)
1 square meter (m^2)	$\approx$	10.763910 square feet (sq ft)
1 square meter (m^2)	$\approx$	1.119599 square yards (sq yd)

Study these formulas of area measure. (A = area)

Circle $\pi = 3.1416$

$A = \pi\, r^2$ r = radius

$A = \dfrac{\pi}{4} d^2$ d = diameter

                    $\dfrac{\pi}{4} = 0.7854$

Rectangle l = length

$A = lw$ w = width

Square s = length of side

$A = s^2$

Trapezoid b_1 = length of base 1

$A = \dfrac{(b_1 + b_2)a}{2}$ b_2 = length of base 2

                    a = altitude (height)

Right Triangle a = altitude

$A = \dfrac{ab}{2}$ b = base

PRACTICAL PROBLEMS

1. Express 0.75 square meter as square millimeters.   _____

2. Express 22,575 square millimeters as square meters.   _____

3. Express 360 square inches as square feet.   _____

4. Find to the nearer thousandth the number of square meters in 2 square feet.   _____

5. Find to the nearer hundredth the number of square meters in 3 square yards.   _____

6. A piece of copper bus bar is 0.75 millimeter by 100 millimeters in cross section. What is the cross-sectional area in square centimeters? _____

7. A square surface cover is 10.16 centimeters on each side. The knockout in the center of the cover has a diameter of 1.27 centimeters. Find the area of the cover to the nearer hundredth centimeter. _____

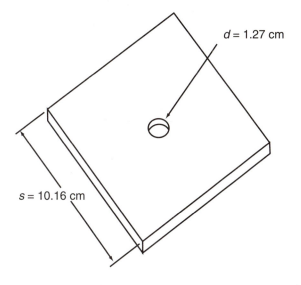

8. An electrician has to pour a concrete pad for a freestanding Federal Pacific 1,600-ampere switch gear. The switch gear measures 75 centimeters wide, 240 centimeters long, and 230 centimeters high. If the pad has to be 30 centimeters larger than the switch gear on all sides, how many meters of surface area will the pad cover? _____

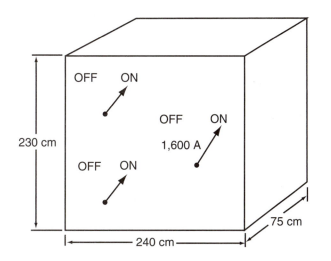

9. A show window requires 15 watts per square meter for sufficient lumination. How many 40-watt fluorescent fixtures are needed for the window shown? Round the answer to the nearer whole number.

_____

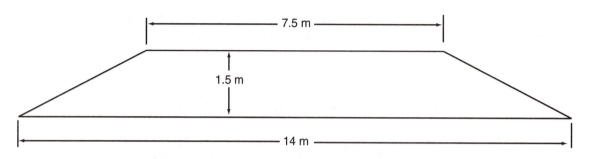

10. A rectangular concrete pad 180 inches wide and 450 inches long supports four transformers. Each transformer base measures 100 inches by 68 inches. What percent of the pad surface area is covered by the transformers? Round the answer to the nearer hundredth percent.

_____

11. A switchboard 3 meters wide and 5 meters long is placed on a concrete slab near a wall. A live part extends 0.15 meter from the back of the switchboard. A clearance of 0.75 meter between the live part and the wall is required. There is a front walk space 1 meter wide and two side walkways each 0.6 meter wide. What is the area of the slab in square meters?

_____

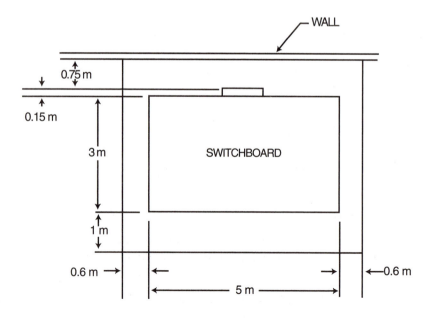

12. A type W4 conductor Neoprene portable cable has an outside diameter of 3.7592 centimeters. What is the cross-sectional area of this cable? Express the answer to the nearer thousandth square centimeter.   _____

13. The area of the floor in a building is 1,525 square feet. The floor is 25 feet wide. One strip of raceway is installed in the center length of the building. Find the number of feet of raceway that are needed.   _____

14. An auditorium is 45 meters wide and 60 meters long. The building code requires 1.2 square meters for each person occupying the auditorium. What is the seating capacity of the auditorium?   _____

15. A conduit has an inside diameter of 3.25 inches. What is the inside area? Round the answer to the nearer thousandth square inch.   _____

16. Capacitance is directly proportional to the surface area of a capacitor's plates. What is the total plate area in square centimeters of a capacitor with 6 plates of the dimension and shape shown? Round the answer to the nearer hundredth square centimeter. (**Note:** A circle has degree measure of 360.)   _____

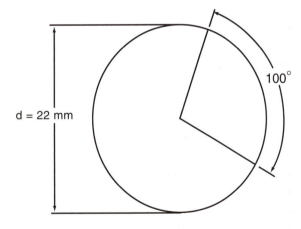

17. The power required to light a store area 60 feet by 20 feet is 5 watts per square foot.   _____

   a. What is the total power requirement?   a. _____

   b. The total load current is found by dividing the power by the voltage. Find the total load current in amperes needed for the area. The voltage is 120 volts.   b. _____

   c. How many 15-ampere circuits are necessary to feed the area?   c. _____

18. An embedded heating cable is installed in a driveway to melt ice and snow. The driveway is 3 meters wide and 40.2 meters long. The cable uses 250 watts of power when activated. How many watts of power are used per square meter? Round the answer to the nearer hundredth.   _____

 # UNIT 28   VOLUME AND MASS MEASURE

BASIC PRINCIPLES OF VOLUME AND MASS MEASURE

Volume Measure

Volume is a three-dimensional measurement. The volume of a box, for example, can be found by multiplying its length, width, and height. The following charts show volume measurements for both the English and metric systems.

VOLUME MEASURE FOR SOLIDS

ENGLISH VOLUME MEASURE FOR SOLIDS	
1 cubic yard (cu yd)	= 27 cubic feet (cu ft)
1 cubic foot (cu ft)	= 1,728 cubic inches (cu in)

METRIC VOLUME MEASURE FOR SOLIDS		
1,000 cubic millimeters (mm^3)	=	1 cubic centimeter (cm^3)
1,000 cubic centimeters (cm^3)	=	1 cubic decimeter (dm^3)
1,000 cubic decimeters (dm^3)	=	1 cubic meter (m^3)
1,000 cubic meters (m^3)	=	1 cubic dekameter (dam^3)
1,000 cubic dekameters (dam^3)	=	1 cubic hectometer (hm^3)
1,000 cubic hectometers (hm^3)	=	1 cubic kilometer (km^3)

ENGLISH–METRIC VOLUME MEASURE FOR SOLIDS		
1 cubic inch (cu in)	≈	16.387064 cubic centimeters (cm^3)
1 cubic foot (cu ft)	≈	0.028317 cubic meter (m^3)
1 cubic yard (cu yd)	≈	0.764555 cubic meter (m^3)
1 cubic centimeter (cm^3)	≈	0.061024 cubic inch (cu in)
1 cubic meter (m^3)	≈	35.314667 cubic feet (cu ft)
1 cubic meter (m^3)	≈	1.037951 cubic yards (cu yd)

VOLUME MEASURE FOR FLUIDS

ENGLISH VOLUME MEASURE FOR FLUIDS	
1 quart (qt)	= 2 pints (pt)
1 gallon (gal)	= 4 quarts (qt)

METRIC VOLUME MEASURE FOR FLUIDS	
10 milliliters (ml)	= 1 centiliter (cl)
10 centiliters (cl)	= 1 deciliter (dl)
10 deciliters (dl)	= 1 liter (l)
10 liters (l)	= 1 dekaliter (dal)
10 dekaliters (dal)	= 1 hectoliter (hl)
10 hectoliters (hl)	= 1 kiloliter (kl)

ENGLISH–METRIC VOLUME MEASURE FOR FLUIDS	
1 gallon (gal)	$\approx$ 3,785.411 cubic centimeters (cm^3)
1 gallon (gal)	$\approx$ 3.785411 liters (l)
1 quart (qt)	$\approx$ 0.946353 liter (l)
1 ounce (oz)	$\approx$ 29.573530 cubic centimeters (cm^3)
1 cubic centimeter (cm^3)	$\approx$ 0.000264 gallon (gal)
1 liter (l)	$\approx$ 0.264172 gallon (gal)
1 liter (l)	$\approx$ 1.056688 quarts (qt)
1 cubic centimeter (cm^3)	$\approx$ 0.033814 ounce (oz)

SOLID-FLUID VOLUME EQUIVALENTS

ENGLISH VOLUME MEASURE EQUIVALENTS	
1 gallon (gal)	= 0.133681 cubic foot (cu ft)
1 gallon (gal)	= 231 cubic inches (cu in)

METRIC VOLUME MEASURE EQUIVALENTS	
1 cubic decimeter (dm^3)	= 1 liter (l)
1,000 cubic centimeters (cm^3)	= 1 liter (l)
1 cubic centimeter (cm^3)	= 1 milliliter (ml)
1,000 milliliters (ml)	= 1 liter (l)

Mass

The mass of an object indicates the amount of matter in the object. Mass and weight are often confused as being the same thing. Mass and weight are proportional only when the object is in a gravitational field. A lead bar on earth could have a weight of 50 pounds, but in outer space it would have no weight. Its mass, however, would be unchanged.

MASS MEASURE

ENGLISH VOLUME MASS MEASURE		
16 ounces (oz)	=	1 pound (lb)
2,000 pounds (lb)	=	1 ton

METRIC MASS MEASURE		
10 milligrams (mg)	=	1 centigram (cg)
10 centigrams (cg)	=	1 decigram (dg)
10 decigrams (dg)	=	1 gram (g)
10 grams (g)	=	1 dekagram (dag)
10 dekagrams (dag)	=	1 hectogram (hg)
10 hectograms (hg)	=	1 kilogram (kg)
1,000 kilograms (Kg)	=	1 megagram (Mg)

ENGLISH–METRIC MASS MEASURE		
1 pound (lb)	≈	0.453592 kilogram (kg)
1 pound (lb)	≈	453.59237 grams (g)
1 ounce (oz)	≈	28.349523 grams (g)
1 ounce (oz)	≈	0.028350 kilogram (kg)
1 kilogram (kg)	≈	2.204623 pounds (lb)
1 gram (g)	≈	0.002205 pound (lb)
1 kilogram (kg)	≈	35.273962 ounces (oz)
1 gram (g)	≈	0.035274 ounce (oz)

PRACTICAL PROBLEMS

1. Find the number of cubic millimeters in 1.2 cubic meters.   _____

2. Find the number of the cubic meters in 2,400 cubic centimeters.   _____

3. Find to the nearer hundredth the number of cubic yards in 42 cubic feet.   _____

4. Find the number of cubic centimeters in 24 cubic millimeters.   _____

5. Express 6 quarts to the nearer hundredth liter.   _____

6. Express 3 gallons to the nearer thousandth liter.   _____

7. Express 9 cubic yards to the nearer hundredth cubic meter.   _____

8. Express 48 ounces to the nearer hundredth liter.   _____

9. Express 5 cubic feet to the nearer thousandth cubic meter.   _____

10. Express 7 cubic inches to the nearer hundredth cubic centimeter.   _____

11. The diameter of an underground fuel tank is 2.6 meters, and it is 7 meters
    long. How many gallons of fuel will the tank hold? Round the answer to the
    nearer gallon.   _____

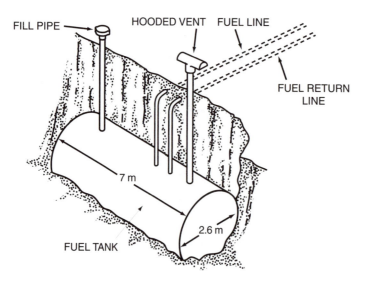

FILL PIPE — HOODED VENT   FUEL LINE

FUEL RETURN
LINE

7 m

2.6 m

FUEL TANK

12.   A power transformer with the capacity of 7 gallons of askarel is one-half full. How many additional liters of askarel are needed to fill the transformer? Round the answer to the nearer tenth.

_____

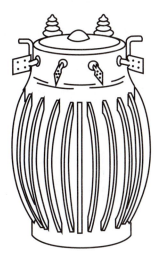

13.   How many cubic meters of air are in a room 10 feet wide, 14 feet long, and 8 feet high? Round the answer to the nearer tenth.

_____

UNIT 29   ENERGY AND TEMPERATURE MEASURE

BASIC PRINCIPLES OF ENERGY AND TEMPERATURE MEASURE

Use the following charts to solve the problems presented in this unit.

MULTIPLICATION FACTOR			PREFIX	SYMBOL
1,000,000	=	10^6	mega	M
1,000	=	10^3	kilo	k
100	=	10^2	hecto	h
10	=	10	deka	da
0.1	=	10^{-1}	deci	d
0.01	=	10^{-2}	centi	c
0.001	=	10^{-3}	milli	m
0.000001	=	10^{-6}	micro	μ
0.000000001	=	10^{-9}	nano	n
0.000000000001	=	10^{-12}	pico	p

QUANTITY	UNIT	SYMBOL
power	kilowatt	kW
	watt	W
electric current	ampere	A
electromotive force	volt	V
electric resistance	ohm	Ω
energy	megajoule	MJ
	kilojoule	kJ
	joule	J
	kilowatt-hour (3.6 MJ)	kW•h
frequency	megahertz	MHz
	kilohertz	kHz
	hertz	Hz
electric capacitance	farad	F
inductance	henry	H

PRACTICAL PROBLEMS

Express the following as numbers in the given units with the proper power of 10.

1. 1.5 amperes
   a. _____ mA
   b. _____ μA

2. 12 volts
   a. _____ mV
   b. _____ μV

3. 2 microfarads
   a. _____ F
   b. _____ pF

4. 3,300 ohms
   a. _____ kΩ
   b. _____ MΩ

5. 37 henries
   a. _____ μH
   b. _____ mH

6. 0.33 kilowatt
   a. _____ W
   b. _____ MW

7. Express 68°F in degrees Celsius.   _____

$$°C = \frac{5}{9}(°F - 32°)$$

8. A 20-ampere fuse element will open at 203°F due to heat from excessive
   current. At what Celsius temperature will the fuse open?   _____

$$°C = \frac{5}{9}(°F - 32°)$$

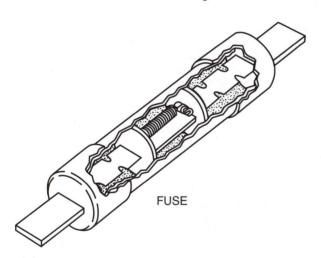

FUSE

9. A coffee percolator rated at 120 volts, 1,400 watts requires 15 minutes to
   complete its cycle. How many megajoules of work are used to make one pot
   of coffee? (3.6 MJ = 1 kW • h)   _____

10. The velocity or ratio wave propagation in free space is 186,000 miles per
    second. What is the velocity in kilometers per second?   _____

11. What is the time required for one alternation in a wave of frequency of 60 hertz? Round the answer to the nearer tenth millisecond.   _____

$$t = \frac{1}{f}$$ where t = time in seconds
f = frequency in hertz
1 = constant (one second)

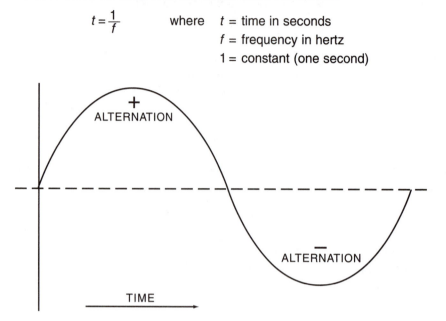

12. The current flow past a given point is 1.758×10^{19} electrons per second. Find to the nearer tenth ampere the current flow past this point.   _____

1 coulomb = 6.28×10^{18} electrons
1 ampere = 1 coulomb/sec

13. Calculate the wavelength in meters for a frequency of 60 hertz.   _____

$$\lambda = \frac{3 \times 10^8}{f}$$ where λ = wavelength
f = frequency in hertz
3×10^8 = a mathematical constant

14. The measured inductance of a power line 110 kilometers long is found to be 295 millihenries.

a. What is the inductance of the line per kilometer? Express the answer to the nearer hundredth millihenry.   a. _____

b. What is the inductance of the line per meter? Express the answer to the nearer hundredth microhenry.   b. _____

 **UNIT 30 COMBINED PROBLEMS ON MEASURE**

This unit provides practical problems involving combined problems on measure.

PRACTICAL PROBLEMS

Express each of the following in the indicated units. Round the answer to the nearer thousandth when necessary.

1. 12 inches as millimeters _____

2. 8 yards as meters _____

3. 5 miles as kilometers _____

4. 10 pints as liters _____

5. 6 gallons as liters _____

6. 20 square feet as square meters _____

7. 15 cubic yards as cubic meters _____

8. 80 pounds as kilograms _____

9. 75 degrees Fahrenheit as degrees Celsius _____

10. 0.740 kilowatt-hour as kilojoules _____

11. 10 kilowatt-hours as megajoules _____

12. 500,000 milliamperes as amperes _____

13. 200 kilovolts as megavolts _____

14. 5,000 microfarads as farads   _____

15. A drum has a radius of 24 inches and a length of 3½ feet. How many liters
     does the drum hold?   _____

16. The diameter of number 10 AWG wire is 0.0285 inch. What is the radius in
     centimeters?   _____

17. The bed of a dump truck measures 5 feet deep, 5 feet wide, and 10 feet long.
     When filled level, how many cubic meters of fill does the truck hold?   _____

18. A power transformer needs 25 gallons of askarel. Find the number of one-liter
     containers of askarel needed for the transformer.   _____

19. The inside diameter of a piece of tubing is 5.5 centimeters, and the outside
     diameter is 6.3 centimeters. What is the thickness of the walls of the tubing?   _____

20. It costs a contractor $90 to drive his pickup truck 500 kilometers. Find the
     cost per kilometer to operate the truck.   _____

21. A bushing 3 centimeters long is cut to a length of 2.25 centimeters. How
     many centimeters are cut off?   _____

22. An office 7 meters by 6 meters is wired for lights. Twenty 40-watt lamps are
     installed. What is the power consumption for the office lighting in watts per
     square meter?   _____

23. A building 50 feet long occupies a surface area of 1,250 feet. Find the width
     of the building.   _____

24. A piece of copper bus bar is 1 centimeter thick by 8 centimeters wide. Find
     the cross-sectional area of the bus bar.   _____

25. A welding cable has an outside diameter of 1.1 millimeters. Find the cross-
     sectional area of this cable.   _____

26. A circular box cover is 10.16 centimeters in diameter and has a 20.42-
     millimeter hole in the center. Find the area of the cover.   _____

Ratio and Proportion

 ## *Unit 31   RATIO*

BASIC PRINCIPLES OF RATIO

Ratio

Ratio is the comparison of two numbers or quantities. Ratios, like fractions, are generally reduced to their lowest terms. The ratio 16:8 (read as 16 to 8) can be reduced to a ratio of 2:1 by dividing both numbers by 8. Ratios also can be written in fractional form. The ratio 5:9 can be written as

$$\frac{5}{9}$$

Inverse Ratios

An *inverse ratio* is the ratio in reverse order of the original ratio. The inverse ratio of 5:9 is 9:5, or

$$\frac{9}{5}$$

Notice that the inverse ratio in fractional form is the reciprocal of the original fraction.

PRACTICAL PROBLEMS

1. Express each ratio in its lowest terms.

    a. 15:5

    b. 25:10

    c. 12:4

    d. ¾:¼

    e. 25:75

    a. _____

    b. _____

    c. _____

    d. _____

    e. _____

2. Find the inverse of each ratio.

    a. 10:3

    b. 5:2

    c. 7:8

    d. ⅞:⅛

    e. 5:12

    a. _____

    b. _____

    c. _____

    d. _____

    e. _____

Note: For problems 3–7, reduce all ratios to their lowest terms.

3. What is the ratio of the number of primary turns to the number of secondary turns in the following diagram?

    _____

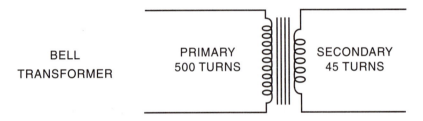

BELL
TRANSFORMER
PRIMARY
500 TURNS
SECONDARY
45 TURNS

4. What is the ratio of the speed of one motor turning at 1,750 revolutions per minute to the speed of a second motor turning at 3,500 revolutions per minute?

    _____

5. What is the ratio of one generator with an output of 3,500 watts to a second generator with an output of 24,500 watts?

    _____

6. If it takes one electrician 18 hours to wire a house and a second electrician 45 hours to wire a similar house, what is the ratio of the second electrician's time to the first electricians time?

    _____

7. What is the ratio of a pinion gear with 14 teeth to a driven gear with 72 teeth?

    _____

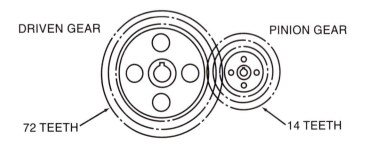

DRIVEN GEAR   PINION GEAR

72 TEETH   14 TEETH

 **UNIT 32   PROPORTION**

BASIC PRINCIPLES OF PROPORTION

Proportion is the equality of two ratios. For example,

$$5:2 = 25:10$$

There are two basic methods of solving problems dealing with proportion. One method is to multiply the *means* together and the *extremes* together. The means are the two inside terms and the extremes are the two outside terms.

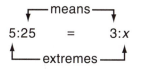

To solve this problem, multiply the extremes together and the means together.

$$5x = 75$$

To find the value of x, divide both sides by 5.

$$x = 15$$

Another method of solving this problem is to use *cross multiplication*. When cross multiplication is used, the two ratios are written as fractions separated by an equal sign.

$$\frac{5}{25} = \frac{3}{x}$$

The top part of one ratio is then multiplied by the bottom part of the other.

$$\frac{5}{25} \diagdown \!\!\!\!\!\diagup \frac{3}{x}$$

$$5x = 75$$

$$x = 15$$

PRACTICAL PROBLEMS

1.   A motor-driven pump discharges 306 gallons of water in 3.6 minutes. How long will it take to discharge 5,200 gallons? Express the answer to the nearer tenth.   _____

2.   If a piece of cable 160 feet long costs $60, what will 500 feet of the same cable cost?   _____

3.   An electrically driven sump pump discharges 125 gallons of water in 4.5 minutes. How much time will it take to discharge 320 gallons?   _____

4.   A copper wire 750 feet long has a resistance of 1.893 ohms. How long is a copper wire of the same diameter whose resistance is 3.156 ohms? Express the answer to the nearer tenth.   _____

5.   A wire whose resistance is 5.075 ohms has a diameter of 31.961 mils. What is the resistance of a wire of the same material and length if the diameter is 40.303 mils? Resistance varies inversely as the square of the diameter. Round the answer to the nearer thousandth.   _____

$$\frac{R_1}{R_2} = \frac{d_2^2}{d_1^2}$$

6.   A wire 2,725 feet long and 85 mils in diameter has a resistance of 0.372 ohm. Find to the nearer thousandth the resistance of 3,600 feet of the same wire.   _____

$$\frac{R_1}{R_2} = \frac{L_1}{L_2}$$

7.   If a wire 1,325 feet long has a resistance of 0.65 ohm, what is the resistance to the nearer hundredth of one mile of the same wire?   _____

8.   Four workers complete a certain electrical job in 120 hours. How long will it take three workers working at the same rate to do the same type of job?   _____

9.   One hundred and twenty feet of conduit cost $35.89. What will 250 feet of conduit cost at the same rate?   _____

10. An electrical repair team of 12 workers completes a job in 288 hours. How long should it take 9 workers working at the same rate to do the same amount of work?   _____

11. If 120 feet of 2-inch conduit cost $154.50, what will 325 feet of 2-inch conduit cost?   _____

12. A 1-foot ruler held perpendicular to the ground casts a shadow 8 inches long. At the same time, a pole casts a shadow 22 feet 4 inches long. What is the height of the pole?   _____

UNIT 33 COMBINED OPERATIONS WITH RATIO AND PROPORTION

This unit provides practical problems involving combined operations with ratio and proportion.

PRACTICAL PROBLEMS

1. A piece of cable 8 feet long costs $60.00. What will 10 feet cost at the same rate? _____

2. A wire 100 feet long has a resistance of 800 ohms. How long is a copper wire of the same area whose resistance is 600 ohms? _____

3. Conduit costs $1,000 for 500 feet. Find the cost of 200 feet at the same rate. _____

4. An electrical repair team of 6 electricians completes a job in 3 hours. How long should it take 2 electricians working at the same rate to do the same amount of work? _____

5. One motor has a speed of 1,800 revolutions per minute. A second motor turns at 3,600 revolutions per minute. What is the ratio of the speed of the first motor to that of the second motor? _____

6. It costs $120 for 10 feet of 4-inch conduit. What will it cost for 50 feet of 4-inch conduit? _____

Note: Use this diagram for problems 7–8.

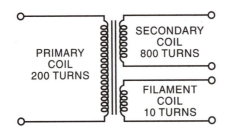

7. Find the ratio of 10 turns of the filament coil to 200 turns of the primary coil. _____

8. Find the ratio of 200 turns of the primary coil to 800 turns of the secondary coil.

_____

9. An electric drill has a chuck attachment to obtain a slower speed. If the attachment has a ratio of 4:1, find the resulting speed if the drill normally turns at 3,600 revolutions per minute.

_____

10. A new "dry pump" is capable of delivering 100 gallons of liquid per minute. How long will it take the pump to deliver 6,000 gallons?

_____

Formulas

Unit 34 REPRESENTATION IN FORMULAS

BASIC PRINCIPLES OF REPRESENTATION IN FORMULAS

A *formula* is a mathematical statement of equality. Just as words are written using symbols, formulas are also written using symbols. In the following problems, written statements will be expressed as mathematical formulas, and formulas will be expressed as written statements.

When writing a formula, it is helpful to know what some of the statements mean and how some of the symbols are used to indicate different operations. The word *sum* is used to indicate addition, and a plus sign (+) is used to represent addition.

Subtraction is generally indicated by statements such as *the difference of*. The minus sign (−) is generally used to represent subtraction.

The word *product* is used to indicate multiplication. There are several ways to indicate that two quantities are to be multiplied together. One way is to simply write two variables (alphabetical characters used to represent the quantities) together with no sign between them. *IR* means to multiply *I* times *R*. Another method is to use parentheses around the values to be multiplied together. *(I)(R)* means to multiply *I* and *R* together. Symbols are also used to represent multiplication. The multiplication sign (×) is often used between numerals but is seldom used with variables. The multiplication dot (•) is often used between two variables to represent multiplication. Another symbol used to represent multiplication, the asterisk (*), has become popular because of computers.

The words *ratio* and *proportional* are represented by division. Division is generally indicated in a formula by writing the numbers as a fraction. If the letter *E* is to be divided by *R,* it would be written as

$$\frac{E}{R}$$

In the formula:

$$I = \frac{E}{R}$$

I is *directly proportional* to *E* and *inversely proportional* to *R*.

Another term often used in mathematics is *reciprocal*. The reciprocal of a number means 1 divided by the number. The reciprocal of 4 is

$$\frac{1}{4}$$

PRACTICAL PROBLEMS

Write a mathematical formula for each statement. Use the symbols given in the problem to write each formula.

1. The total resistance (R_t) of a series electrical circuit is equal to the sum of the individual resistances $(R_1, R_2, \ldots, R_n)$.   _____

2. The total capacitance (C) is directly proportional to the product of the dielectric constant (k) and plate area (A) and inversely proportional to the thickness (d) of the dielectric.   _____

3. In a power transformer, the ratio of the primary voltage (E_p) to the secondary voltage (E_s) is equal to the ratio of the power in the primary (P_p) to the power in the secondary (P_s).   _____

4. The current (I) in an electrical circuit is directly proportional to the voltage applied (E_a) and inversely proportional to the circuit impedance (Z).   _____

5. The capacitance (C) may be found by taking the reciprocal of the product of two pi (2π), the frequency (f), and the capacitive reactance (X_c).   _____

6. An electrical motor has a degree of efficiency (Eff) equal to the ratio of its useful output power (P_o) to the input power (P_i) required to operate it.   _____

7. The electrical power (P) dissipated by an incandescent lamp is directly proportional to the square of the voltage drop (E) across the lamp and inversely proportional to the electrical resistance (R) of the lamp.   _____

8. The electrostatic force *(F)* between two plates of a charged capacitor is directly proportional to the product of the charges (q_1 and q_2) on each plate and inversely proportional to the square of the distance *(d)* between the plates.   _____

9. Between two coils there exists a mutual inductance (L_m) that is equal to the square root of the product of the two inductances (L_1 and L_2) multiplied by the coefficient of coupling *(k)* between the two coils.   _____

10. The resonant frequency *(f_r)* of a circuit containing both inductance *(L)* and capacitance *(C)* may be found by the reciprocal of two pi (2π) multiplied by the square root of the product of the inductance and capacitance.   _____

Write a statement for each of the formulas given.

11. $G = \dfrac{1}{R}$ where G = conductance
R = resistance

_____

_____

12. $R_s = \dfrac{I_m R_m}{I_s}$ where R_s = resistance of the ammeter shunt
I_m = current through the meter movement
R_m = resistance of the meter movement
I_s = current through the shunt

_____

_____

_____

13. $E = 0.707E_{max}$ where E =   effective value of AC voltage

E_{max} =   maximum peak value of AC voltage

0.707   =   a derived constant

---

14. $R_s = \dfrac{R_t}{Z}$ where R_s =   power factor of an AC circuit

R_t =   total circuit resistance

Z =   circuit impedance

---

15. $f = \dfrac{PN}{60}$ where f =   frequency of the generated AC voltage, in hertz

P =   number of pairs of poles of the generator

N =   revolutions per minute of the generator in the magnetic field

60   =   a constant representing the number of seconds in a minute

UNIT 35   REARRANGEMENT IN FORMULAS

BASIC PRINCIPLES OF REARRANGING FORMULAS

A mathematical formula is generally referred to as an *equation*. An equation is a statement of equality that indicates that the quantity on one side of the equal sign is equal to the quantity on the other side of the equal sign. A pair of balanced scales is a good example of this principle. As long as the weight is the same on each side of the scales, they will be in balance. Weight can be added to or removed from the scales without affecting the balance as long as the same amount of weight is added to or removed from both sides.

The same principle is true for an equation. Any operation can be performed on one side of the equation as long as the same operation is performed on the other side. In the following problems, formulas will be rearranged to find different quantities. This is done by rearranging the formula so that the quantity to be found is on one side of the equal sign by itself.

Example: If $X = 2JLC$, solve for L.

To solve for L means to get L on one side of the equal sign by itself. This can be done by removing all the other factors on the right side of the equation. The formula states that X is equal to 2 times J times L times C. Multiplication is the operation indicated. To remove all the factors except L, perform the inverse operation, which is division:

$$\frac{X}{2JC} = \frac{2JLC}{2JC}$$

Notice that both sides of the equation are divided by $2JC$.

$$\frac{X}{2JC} = \frac{2J\!\!\!/ L C\!\!\!/}{\underset{1}{2J\!\!\!/ C\!\!\!/}}$$

The equation is now:

$$\frac{X}{2JC} = \frac{L}{1}$$

The formula can now be written:

$$L = \frac{X}{2JC}$$

Notice that $\frac{L}{1}$ is the same as L.

When rearranging formulas, remember two principles:

1. Anything can be done to one side of an equation as long as the same thing is done to the other side.

2. When removing a variable or number from one side of the equation, perform the inverse operation of the one that is indicated.

PRACTICAL PROBLEMS

Solve each of the equations for the variable shown.

1. $Q = CV$, solve for C.   _____

2. $I = \dfrac{E}{Z}$, solve for Z.   _____

3. $R^2 = Z^2 - X^2$, solve for Z.   _____

4. $R\dfrac{P}{I^2}$, solve for I.   _____

5. $R_2 = R_t - R_1 - R_3$, solve for R_t.   _____

6. $X_L = 2\pi fL$, solve for L.   _____

7. $X_c = \dfrac{1}{2\pi fC}$, solve for C.   _____

8. $pf = \dfrac{R}{X}$, solve for R.   _____

9. $N_s = \dfrac{E_s N_p}{X}$, solve for N_p.   _____

10. $P = \dfrac{120f}{N}$, solve for f.   _____

11. $\dfrac{Z_P}{Z_s} = \dfrac{N_P^2}{N_s}$, solve for Z_P.   _____

12. $E_s I_s = E_p I_p$, solve for E_p.   _____

13. $u = g_m r_p$, solve for r_p.   _____

14. $C = \dfrac{(0.088)(4)kA(N-1)}{N}$, solve for A.   _____

15. $L_m = k\sqrt{L_1 L_2}$, solve for k.   _____

UNIT 36   GENERAL SIMPLE FORMULAS

BASIC PRINCIPLES OF SIMPLE FORMULAS

To solve a formula, the variables are replaced with numbers that are the known values of the variables. Multiplication, division, addition, and subtraction are then performed as indicated by the formula.

Example: An electric iron has a current draw of 5 amperes and a resistance of 10 ohms. How much power is being consumed by this iron? The formula for this problem is $P = I^2R$ where

$$P = \text{power in watts}$$
$$I = \text{current in amperes}$$
$$R = \text{resistance in ohms}$$

To solve the problem, substitute number values for the letter values in the formula.

$$P = 5 \times 5 \times 10$$
$$P = 250 \text{ watts}$$

PRACTICAL PROBLEMS

1. Find the total resistance (R_t) of the three field rheostats shown. Express the
    answer to the nearer hundredth.   _____

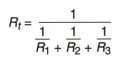

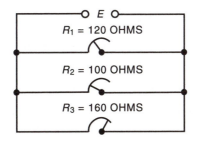

2. A motor takes 38 amperes on a 220-volt circuit. Find the horsepower output *(hp)* of the motor shown with an efficiency of 90%. Express the answer to the nearer hundredth.

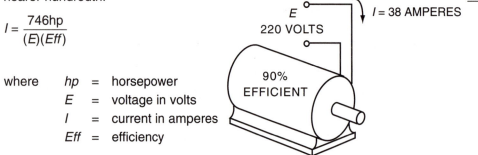

$$I = \frac{746hp}{(E)(Eff)}$$

where   *hp* = horsepower
         E   = voltage in volts
         I   = current in amperes
         Eff = efficiency

3. A circuit uses 124 amperes at a voltage of 230 volts. Find the load *(P)* in kilowatts.

$$P = \frac{EI}{1,000}$$

where   *P* = power in kilowatts
         E = voltage in volts
         I   = current in amperes

4. Find the resistance of 240 feet of number 10 copper wire with a cross-sectional area *(d²)* of 10,380 circular mils. Express the answer to the nearer hundredth.

$$R = \frac{KL}{d^2}$$

where   *R* = resistance in ohms
         K = specific resistance of copper (10.8)
         L = length in feet
         d = diameter in mils

5. A cell has a voltage of 2.4 volts and an internal resistance of 0.03 ohm. The cell is connected to an electromagnet with a resistance of 1.5 ohms. What current, to the nearer hundredth, does the magnet receive?

$$I = \frac{E}{R + r}$$

where   *I* = current in amperes
         E = voltage in volts
         r = internal resistance in ohms
         R = resistance of external circuit in ohms

6. What is the resistance of R_1 if the total resistance (R_t) of the parallel circuit shown is 3.2 ohms? _____

$$R_t = \frac{1}{\frac{1}{R_1} + \frac{1}{R_2} + \frac{1}{R_3} + \frac{1}{R_4}}$$

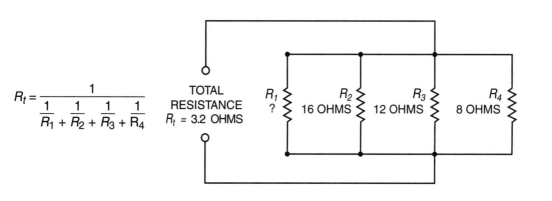

TOTAL RESISTANCE
R_t = 3.2 OHMS

R_1 ? R_2 16 OHMS R_3 12 OHMS R_4 8 OHMS

7. The impressed voltage across a motor armature is 230 volts, and the armature resistance is 0.48 ohm. The countervoltage equals 224 volts. What current does the armature take? _____

$$I = \frac{E_x - E_c}{R}$$

where I = current in amperes
   R = resistance in ohms
   E_x = impressed voltage in volts
   E_c = countervoltage in volts

8. The internal resistance of five cells is 1.8 ohms each. The cells are connected as shown in the diagram to a 2-ohm resistance. What current, to the nearer hundredth, exists in the circuit? _____

$$I = \frac{(E)(ns)}{(r)(ns) + R}$$

where E = voltage of one cell in volts
   ns = number of cells in series
   r = internal resistance of one cell in ohms
   R = resistance of external circuit in ohms
   I = current in amperes

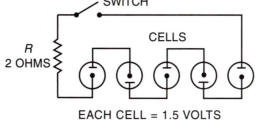

SWITCH

CELLS

R
2 OHMS

EACH CELL = 1.5 VOLTS

Note: Use this formula for problems 9–11.

$$°F = \frac{9}{5}(°C) + 32°$$ where $°C$ = temperature in degrees Celsius
   $°F$ = temperature in degrees Fahrenheit

9.   The temperature of an oven is 200°C. Find the temperature of the oven in degrees Fahrenheit.

_____

10.   The temperature of a motor rises to 50°C above room temperature. The room temperature is 26°C. What will a Fahrenheit thermometer read?

_____

11.   If the temperature of a commutator is 65°C, what is the temperature in degrees Fahrenheit?

_____

12.   If R_x is the unknown resistance of the Wheatstone bridge arrangement shown in the figure, find the value of this unknown resistance.

_____

$$\frac{R_x}{R_3} = \frac{R_1}{R_2}$$

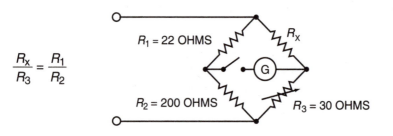

R_1 = 22 OHMS R_x

R_2 = 200 OHMS R_3 = 30 OHMS

13.   The voltage drop across a 100-ohm standard resistance is 16 volts. What is the resistance of a coil connected in series if the voltage drop across the coil is 8 volts?

_____

$$\frac{V_R}{R_x} = \frac{R}{R_x}$$

where V_R = voltage drop across standard resistance
   V_x = voltage drop across unknown resistance
   R = resistance of standard in ohms
   R_x = resistance of coil in ohms

14. Determine the horsepower output to the nearer hundredth of the motor in the circuit shown. _____

$$hp\ input = \frac{EI - 2RI^2}{746}$$

where

E = voltage in volts
I = current in amperes
R = resistance in ohms

hp output = hp input × efficiency

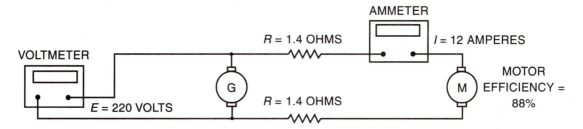

VOLTMETER

$R = 1.4$ OHMS

AMMETER

$I = 12$ AMPERES

$E = 220$ VOLTS

G

$R = 1.4$ OHMS

M

MOTOR
EFFICIENCY =
88%

15. A motor turns 1,130 revolutions per minute (r/min) with a 6-inch driving pulley *(d)*. What are the revolutions per minute of a 24-inch pulley *(D)* belted to this motor? _____

$$r/min\ of\ D = \frac{Diameter\ of\ d \times r/min\ of\ d}{Diameter\ of\ D}$$

16. Find the power used in the single-phase circuit shown. _____

$P = (E)(I)(PF)$

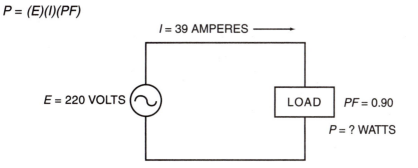

$I = 39$ AMPERES ⟶

$E = 220$ VOLTS

LOAD $PF = 0.90$

$P = ?$ WATTS

17. A 2-phase, 4-wire, 220-volt circuit has a current of 16.5 amperes per leg. The power factor is 0.85. Find the number of watts consumed. For a 2-phase, 4-wire circuit: _____

$$P = 2(E)(I)(PF)$$

18. A motor takes a current of 27.5 amperes per leg on a 440-volt, 3-phase circuit. The power factor is 0.80. What is the load in watts? Round the answer to the nearer whole watt. For a 3-phase circuit: _____

$$P = \sqrt{3}\ (E)(I)(PF)$$

19. What is the current per leg in a 2-phase, 3-wire circuit if the power is 200 watts, the voltage is 120 volts, and the power factor is 0.9? Express the answer to the nearer thousandth. For a 2-phase circuit: _____

Note: Use this formula for problems 20–22.

	where		
	I	=	current in amperes _____
	E	=	voltage in volts _____
$hp = \dfrac{(E)(I)(Eff)}{746}$	Eff	=	efficiency of motor _____
	hp	=	horsepower _____

20. Solve the equation for the current *(I)* required by a motor when the horsepower, percent efficiency of the motor, and the voltage are known.   _____

21. A 12-horsepower motor is connected to a 220-volt circuit. How much current, to the nearer hundredth, does it take if the motor efficiency is 87%?   _____

22. If a 25-horsepower motor takes 96 amperes at full load, and the motor efficiency is 90%, what is the terminal voltage? Express the answer to the nearer hundredth.   _____

Note: Use this formula for problems 23–24.

	where		
$Z = \sqrt{R^2 + X^2}$	Z	=	impedance in ohms _____
	R	=	resistance in ohms _____
	X	=	reactance in ohms _____

23. What is the impedance to the nearer hundredth of the circuit shown in the figure? _____

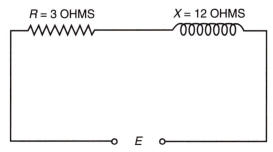

$R = 3$ OHMS $X = 12$ OHMS

E

24. Find the impedance of a coil with 5.5 ohms resistance and 8.2 ohms reactance. Round the answer to the nearer hundredth. _____

Note: Use this formula for problems 25–26.

$$Z = \frac{E}{I}$$

where Z = impedance in ohms
E = voltage in volts
I = current in amperes

25. A circuit has a voltage of 440 volts across it and a current of 5.5 amperes. What is the impedance of the circuit? _____

26. If a current of 4.5 amperes exists, what is the impedance of a circuit having a voltage of 110 volts across it? Express the answer to the nearer hundredth. _____

Note: Use this formula for problems 27–28.

$$kW\ load = \frac{\sqrt{3}(E)(I)(PF)}{1,000}$$

where $kW\ load$ = 3-phase power in kilowatts
I = current in amperes
E = voltage in volts
PF = power factor

27. A 3-phase, 3-wire circuit supplying power to several motors delivers 275 amperes. If the voltage is 220 volts and the power factor is 0.80, what is the kilowatt load? Round the answer to the nearer hundredth.   _____

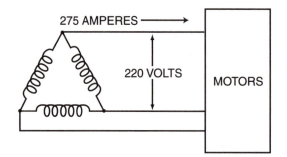

28. A 3-phase, 3-wire circuit delivers 175 amperes at a voltage of 220 volts with a 0.85 power factor. What is the kilowatt load to the nearer hundredth?   _____

Note: Use this formula for problem 29.

$$kW\ load = \frac{2(E)(I)(PF)}{1,000}$$

where   kW load   =   2-phase power in kilowatts
I =   current in amperes
E =   voltage in volts
PF =   power factor

29. A 2-phase, 4-wire circuit with a voltage of 220 volts is connected to a group of motors that requires 85 amperes with a 0.80 power factor. What is the load in kilowatts?   _____

30. What is the current in amperes in a 3-phase circuit when the load is 67.5 horsepower, the efficiency is 90%, the voltage is 440 volts, and the circuit power factor is 0.80? Round the answer to the nearer hundredth.   _____

$$hp = \frac{\sqrt{3}(E)(I)(Eff)(PF)}{746}$$

31. A 2-wire circuit supplying power to a load 180 feet from the source has a 6.5 voltage drop. If the current is 37.5 amperes, what to the nearer hundredth is the cross-sectional area *(CM)* of the wire?   _____

$$CM = \frac{KNLI}{E_e}$$

where

CM	=	cross-sectional area of conductor in circular mils
L	=	length of one wire in feet
K	=	constant for copper (10.8)
I	=	current in amperes
E_e	=	voltage drop in volts
N	=	number of wires

32. In a circuit, what current does a 220-volt, 5-horsepower motor receive if the efficiency of the motor is 80%? Express the answer to the nearer hundredth.   _____

$$hp = \frac{(E)(I)(Eff)}{746}$$

33. In a 2-wire circuit, what size conductors to the nearer hundredth does a 220-volt, 5 horsepower motor require if the voltage drop is 3 volts and the distance is 160 feet from the meter? The motor has an 80% efficiency.   _____

$$CM = \frac{746 \, hpKNL}{E_e E_m Eff}$$

where

CM	=	circular mil area of conductor
hp	=	horsepower
L	=	length of line wires in feet
E_e	=	voltage drop
K	=	constant for copper (10.8)
Eff	=	efficiency of motor
E_m	=	motor voltage
N	=	number of wires

34. What size conductor should be used in a 2-wire system to service a load of 6,500 watts, 150 feet from the generator? The voltage at the load is 120 volts, and the voltage drop should not exceed 2.5 volts. Express the answer to the nearer hundredth.   _____

$$I = \frac{P}{E} \qquad CM = \frac{KNLI}{E_e}$$

35. Find the total voltage *(E)* for the circuit shown. Express the answer to the nearer tenth.

_____

$$E = \sqrt{E_R^2 + (E_L - E_C)^2}$$

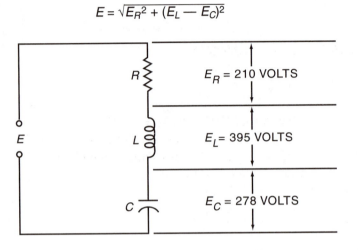

E_R = 210 VOLTS

E_L = 395 VOLTS

E_C = 278 VOLTS

36. What is the frequency in hertz (Hz) of the current furnished by an alternator having 8 poles and running at a speed of 900 revolutions per minute?

_____

$$f = \frac{PN}{60}$$

where f = frequency in hertz

P = number of pairs of poles

N = revolutions per minute

UNIT 37   OHM'S LAW FORMULAS

BASIC PRINCIPLES OF OHM'S LAW FORMULAS

The formula for Ohm's law is

$$E = IR$$

where
E = voltage in volts
I = current in amperes
R = resistance in ohms

This formula can be rearranged and solved for *hl* to give

$$I = \frac{E}{R}$$

or for *R* to give

$$R = \frac{E}{I}$$

To solve problems using Ohm's law, substitute numerical values for variables in the formulas. It will be necessary to choose the proper formula. This is done by determining which quantity is to be found and which quantities are known.

Example: An electric toaster has a resistance of 10 ohms and draws a current of 12 amperes when connected to the line. To what voltage is the toaster connected?

The value to be found is voltage *(E)*. The values known are resistance *(R)* and current *(I)*. The formula chosen must contain these three quantities. The formula *E = IR* will be used with the values *I* = 12 amperes and *R* = 10 ohms substituted.

$$E = IR$$
$$E = 12 \times 10$$
$$E = 120 \text{ volts}$$

PRACTICAL PROBLEMS

Express the answer to the nearer hundredth, when necessary.

1.   How much current *(I)* flows through a lamp that has a resistance *(R)* of 220 ohms and is connected across a 120-volt *(E)* circuit?   _____

2.   What voltage is required to force 4.5 amperes of current through an electric iron having a resistance of 25.5 ohms?   _____

Note: Use this diagram and the formulas given for problems 3–4.

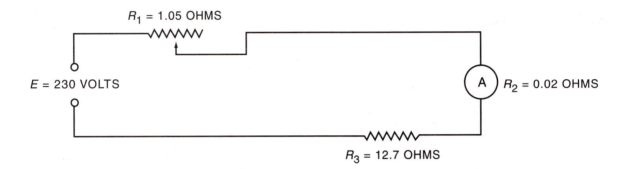

$R_1 = 1.05$ OHMS

$E = 230$ VOLTS

A $R_2 = 0.02$ OHMS

$R_3 = 12.7$ OHMS

$$R_T = R_1 + R_2 + R_3$$
$$E = IR_T$$

3. What current flows through the circuit shown in the diagram?   _____

4. Find the value of the current for the circuit if $R_1 = 0$.   _____

5. An electric heater radiates sufficient heat at a 10-ampere load on a 120-volt circuit. What is the hot resistance?   _____

6. A lamp with a resistance (hot) of 50 ohms is connected across 120 volts. What current does the lamp receive?   _____

7. The 2.4-volt cell with an internal resistance of 0.5 ohm is connected to an external resistance of 1.25 ohms. What current does the external resistance (R_L) receive?   _____

$R = r + R_L$ where r =   internal resistance
                                  R_L =   external resistance

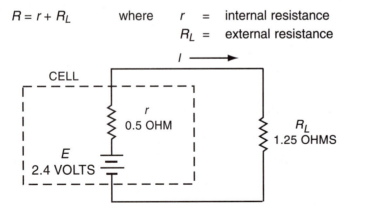

CELL

r
0.5 OHM

R_L
1.25 OHMS

E
2.4 VOLTS

8. A lamp requires a current of 0.94 ampere when connected to a circuit with a voltage of 120 volts. Find the resistance in ohms of the lamp.   _____

Note: Use this diagram for problems 9–10.

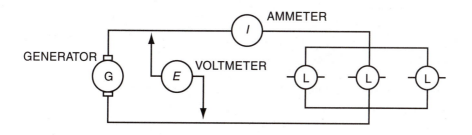

9. In the circuit shown, three lamps have a total resistance of 220 ohms. The voltmeter reading *(E)* is 120 volts. Find the ammeter reading *(I)* in this circuit.   _____

10. Find the total resistance in ohms if the current in the circuit is 2.82 amperes and the voltage is 120 volts.   _____

11. The total resistance of the two line wires is 0.314 ohm. The cross-sectional area of the wire, which equals d^2, is 16,510 circular mils. What is the length of each wire?   _____

$$R = \frac{KL}{d^2}$$

where
R = resistance in ohms
K = 10.8 (constant for copper)
L = length of wire in feet
d = diameter in mils

12. If the cross-sectional area of a wire is 16,510 circular mils, what is the diameter *(d)* of the wire?   _____

Cross-sectional area = d^2 where d = diameter in mils   _____

Note: Use this formula for problems 13–14.

$$R = \frac{KL}{d^2}$$

where R = resistance in ohms
   L = length in feet
   d = diameter in mils
   K = 10.8 (constant)

13. What is the length of a wire having a resistance of 1.085 ohms and a diameter of 162.02 mils?   _____

14. Find the length of a copper wire 0.104 inch in diameter if it has a resistance of 3.1 ohms. (1 mil = 0.001 inch)   _____

15. What is the coil resistance of an electric bell if 0.25 ampere exists when a voltage of 2.8 volts is applied?   _____

16. The four dry cells are connected in series as shown.   _____

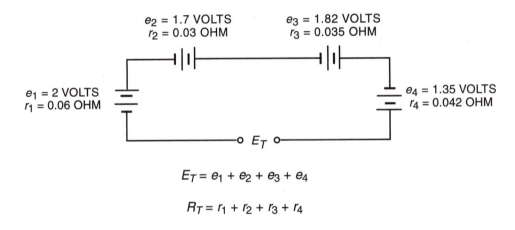

$e_2 = 1.7$ VOLTS
$r_2 = 0.03$ OHM

$e_3 = 1.82$ VOLTS
$r_3 = 0.035$ OHM

$e_1 = 2$ VOLTS
$r_1 = 0.06$ OHM

$e_4 = 1.35$ VOLTS
$r_4 = 0.042$ OHM

E_T

$$E_T = e_1 + e_2 + e_3 + e_4$$

$$R_T = r_1 + r_2 + r_3 + r_4$$

   a. Find the total resistance.   a. _____
   b. Find the total voltage.   b. _____

17. Using Ohm's law, find the resistance of one conductor of an annunciator cable if the voltage across the conductor is 1.2 volts and the current is 2 amperes.   _____

18. A 60-volt electric time clock circuit with several clocks connected in parallel has a 2-wire line voltage drop of 12.5 volts when the circuit is closed. Using Ohm's law, find the total resistance of the 2-wire line if a current of 10 amperes exists.   _____

19. What is the resistance of an electromagnet connected to a cell having a voltage of 2.2 volts and drawing a current of 0.3 ampere? _____

20. Find the total resistance *(R$_T$)* of the pictured set-up. _____

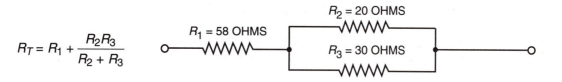

$$R_T = R_1 + \frac{R_2 R_3}{R_2 + R_3}$$

$R_1 = 58$ OHMS

$R_2 = 20$ OHMS

$R_3 = 30$ OHMS

21. Eight resistors are connected as shown. The value of each resistor is 0.5 ohm. Find the total resistance *(R$_T$)*. _____

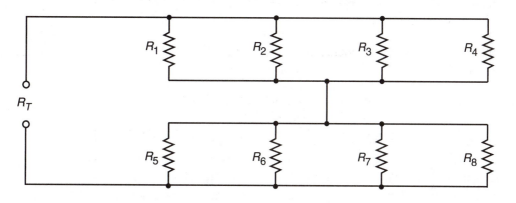

$$R_T = \frac{1}{\dfrac{1}{R_1} + \dfrac{1}{R_2} + \dfrac{1}{R_3} + \dfrac{1}{R_4}} + \frac{1}{\dfrac{1}{R_5} + \dfrac{1}{R_6} + \dfrac{1}{R_7} + \dfrac{1}{R_8}}$$

22. Find the total resistance *(R$_T$)* of the three low-voltage relay coils. Round the answer to the nearer hundredth. _____

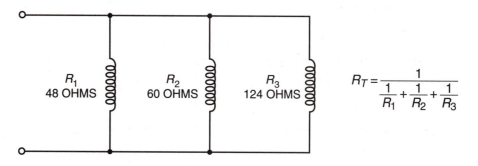

R_1 48 OHMS

R_2 60 OHMS

R_3 124 OHMS

$$R_T = \frac{1}{\dfrac{1}{R_1} + \dfrac{1}{R_2} + \dfrac{1}{R_3}}$$

23. A wire has a cross-sectional area of 10,380 circular mils. Find the resistance of 900 feet of wire supplying charging current for a 24-volt battery. Express the answer to the nearer thousandth.   _____

$$R = \frac{KL}{CM}$$

where   CM = cross-sectional area in circular mils
         K = 10.8
         L = total length of wire in feet
         R = resistance in ohms

24. One 2-wire section of a 24-volt watchman clock signal system is operated through a 450-foot length of lead-covered cable. The size of the conductor is number 18, and the diameter is 40.3 mils. What is the resistance of one conductor? Express the answer to the nearer thousandth.   _____

$$R = \frac{KLN}{d^2}$$

where   N = number of wires
         K = constant (10.8)
         L = length of one wire in feet
         R = resistance in ohms
         d = diameter in mils

25. A 30-conductor control cable 2,150 feet long is installed between the engine room and a small substation. The voltage drop is 5.5 volts, and the line current is 12 amperes for one 2-wire circuit during operation. What is the cross-sectional area of the wire in circular mils (CM) for one 2-wire circuit?   _____

$$CM = \frac{KLNI}{E_e}$$

where   N = number of wires
         E_e = voltage drop
         L = length in feet per wire
         CM = cross-sectional area in circular mils
         I = current in amperes
         K = 10.8

UNIT 38   POWER FORMULAS

BASIC PRINCIPLES OF POWER FORMULAS

The following problems will be solved using the power formulas and Ohm's law.

$$P = EI \qquad I = \frac{E}{R} \qquad E = IR \qquad R = \frac{E}{I}$$

$$P = I^2R \qquad I = \frac{P}{E} \qquad E = \frac{P}{I} \qquad R = \frac{E^2}{P}$$

$$P = \frac{E^2}{R} \qquad I = \sqrt{\frac{P}{R}} \qquad E = \sqrt{PR} \qquad R = \frac{P}{I^2}$$

PRACTICAL PROBLEMS

1. A circuit is connected to a voltage *(E)* of 240 volts and has a current *(I)* of 15 amperes. How many watts of power are used?   _____

2. A 500-watt electric iron is connected to 120 volts. How many amperes of current to the nearer hundredth does the iron take?   _____

3. Find the current to the nearer thousandth ampere used by the lamp shown.   _____

E
120 VOLTS

75-WATT
LAMP

4. A one-horsepower motor uses 900 watts when connected to a circuit. The motor receives a current of 3.4 amperes. Find the voltage to the nearer tenth across the circuit.   _____

5. A heater rated at 7.5 amperes is connected to a 240-volt circuit. Find the power in watts for the heater.   _____

6. How many watts will a motor use if the voltage is 240 volts and the motor is drawing 40 amperes of current?   _____

7. What voltage to the nearer hundredth volt is necessary to create a current of 1.74 amperes in a 200-watt lamp?   _____

8. A circuit has a voltage of 120 volts and a current of 7.5 amperes. Find the amount of power used.   _____

9. What voltage to the nearer hundredth volt is necessary for a circuit of 1,625 watts with a current of 14.3 amperes?   _____

10. A 550-watt electric iron has a voltage drop of 112 volts. What current to the nearer hundredth ampere exists in the iron?   _____

11. The resistance coil of the 400-watt soldering iron is shown. Find the current to the nearer hundredth ampere received by the resistance coil of the soldering iron.   _____

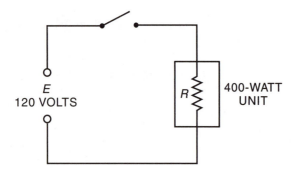

12. A circuit uses 216.2 watts and draws a current of 1.88 amperes. Find the resistance. Express the answer to the nearer hundredth ohm.   _____

$$E = \frac{P}{I} \quad \text{and} \quad R = \frac{E}{I}$$

13. What is the total resistance in a circuit having a voltage of 120 volts and a power of 500 watts? Express the answer to the nearer tenth ohm.   _____

$$E = \frac{P}{I} \quad \text{and} \quad R = \frac{E}{I}$$

14. A motor uses 5,595 watts. Find the horsepower load of the motor.   _____

$$hp = \frac{P}{746}$$

15. What is the power in watts in a circuit drawing 59 amperes of current with a total line resistance of 0.67 ohm? _____

$$P = I^2 R$$

16. Find the number of watts in the armature of a generator having an internal resistance of 0.3 ohm and a voltage of 6 volts across the resistance. _____

$$P = \frac{E^2}{R}$$

UNIT 39   COMBINED PROBLEMS ON FORMULAS

This unit provides practical problems involving combined problems on formulas.

PRACTICAL PROBLEMS

$$R = \frac{E}{I} \qquad\qquad I = \frac{E}{R} \qquad\qquad E = IR$$

1. How much current will exist in a lamp that has a resistance of 100 ohms and is connected across a 240-volt circuit? _____

2. What voltage will be required to create a current of 5 milliamperes through a wire-wound resistor having a resistance of 750 ohms? _____

3. An electric heater radiates sufficient heat with a 12-ampere load on a 240-volt circuit. What is the hot resistance? _____

4. What voltage is necessary if 1.5 amperes exist in a 60-watt lamp? _____

$$P = EI$$

5. A circuit is connected to a voltage of 120 volts and has a current of 8 milliamperes. How many watts are used? _____

6. A 1,500-watt electric iron is connected to 120 volts. How many amperes will the iron take? _____

7. What is the length of a copper wire 5 mils in diameter, having a resistance of 13.6 ohms? Round the answer to the nearer tenth. _____

$$R = \frac{KL}{d^2}$$

where
R = resistance in ohms
L = length in feet
d = diameter in mils
K = 10.8

8. What current will exist in a 3,000-watt electric iron when a voltage of 240 volts is maintained? _____

$$I = \frac{P}{E}$$

9. What is the current in a circuit having a resistance of 5,000 ohms if it is consuming 1,500 watts? Round the answer to the nearer thousandth. _____

$$I = \sqrt{\frac{P}{R}}$$

10. A cell has a voltage *(E)* of 2.1 volts, an internal resistance *(r)* of 0.9 ohm, and is connected to an electromagnet with a resistance *(R)* of 0.3 ohm. What current, in amperes, will the electromagnet receive? _____

$$I = \frac{E}{R + r}$$

11. The impressed voltage *(E_x)* across a motor armature is 440 volts, and the armature resistance is 100 ohms. If the countervoltage *(E_c)* equals 10 volts, how many amperes does the armature take? _____

$$I = \frac{E_x - E_c}{R}$$

12. The temperature of a commutator is 20°C. Express this temperature in degrees Fahrenheit. _____

$$°F = \frac{°C}{\frac{5}{9}} + 32$$

13. What is the amount of wattage used in a single-phase circuit drawing 7.2 amperes at 120 volts with a 0.8 power factor? _____

$$P = (E)(I)(PF)$$

14. A motor takes a current of 8 amperes per leg on a 440-volt, 3-phase circuit. The power factor is 0.75. What is the load in watts? _____

$$P = \sqrt{3}(E)(I)(PF)$$

15. If a 5-horsepower motor running at full load is connected to a 240-volt circuit, how much current will it take if the motor efficiency is 80%? Round the answer to the nearer thousandth.

_____

$$hp = \frac{(E)(I)(Eff)}{746}$$

16. What is the impedance (Z) of a coil with 84 ohms resistance (R) and 500 ohms reactance (X)? Round the answer to the nearer ohm.

_____

$$Z = \sqrt{R^2 + X^2}$$

17. What is the impedance of a circuit having a voltage of 120 volts across it and a current of 0.025 ampere through it?

_____

$$Z = \frac{E}{I}$$

Trigonometry / SECTION 9

 ## Unit 40   RIGHT TRIANGLES

BASIC PRINCIPLES OF RIGHT TRIANGLES

The *Pythagorean theorem* was developed by a Greek mathematician named Pythagoras almost three thousand years ago. It is a basic law that describes the relationship of the sides of a *right triangle* (a triangle that contains a right, or 90°, angle). The Pythagorean theorem states that the **sum of the squares of the sides of a right triangle is equal to the square of the hypotenuse**. The hypotenuse is the longest side of a right triangle and is located opposite the right or 90° angle.

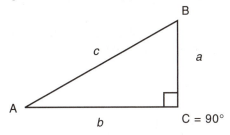

Using the letter c to represent the length of the hypotenuse and the letters a and b to represent the two sides, formulas can be derived to find the length of any side when two others are known using the formula:

$$a^2 + b^2 = c^2$$

The formula for the theorem can be rearranged to give the following equations:

$$c^2 = a^2 + b^2 \qquad c = \sqrt{a^2 + b^2}$$

$$a^2 = c^2 - b^2 \qquad a = \sqrt{c^2 - b^2}$$

$$b^2 = c^2 - a^2 \qquad b = \sqrt{c^2 - a^2}$$

Example: The hypotenuse of a right triangle is 42 feet long. The base is 28 feet in length. Calculate the altitude of the triangle.

$$a = \sqrt{c^2 - b^2} = \sqrt{(42)^2 - (28)^2} = \sqrt{1{,}764 - 784} = \sqrt{980} = 31.30 \text{ inches}$$

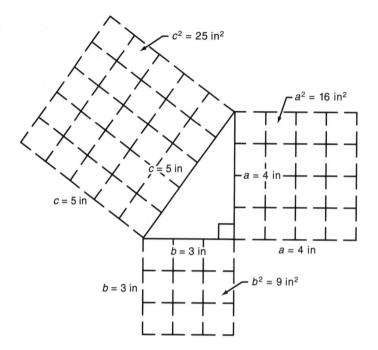

Figure 1. The area of the square connected to the base plus the area of the square connected to the altitude is equal to the area of the square connected to the hypotenuse.

The triangle shown in Figure 1 is a right triangle with a base 3 inches long, altitude 4 inches long, and hypotenuse 5 inches long. Attached to each side is a square. The base square is $3 \times 3 = 9$ *square* inches. The square using the altitude as one of its sides is $4 \times 4 = 16$ *square* inches. The square on the hypotenuse $5 \times 5 = 25$ *square* inches. Interestingly, the sum of the base and altitude areas (9 in² + 16 in²) equals the square of the hypotenuse (25 in²). This can be written $c^2 = a^2 + b^2$.

PRACTICAL PROBLEMS

Express the answer to the nearer hundredth, when necessary.

Note: Use this illustration for problems 1–6.

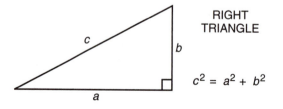

1. $a = 12$ $b = 15$ Find c.   _____

2. $a = 24$ $c = 36$ Find b.   _____

3. $c = 60$ $b = 30$ Find a.   _____

4. $a = 28$ $b = 18$ Find c.   _____

5. $c = 48.25$ $b = 22.75$ Find a.   _____

6. $a = 12\frac{3}{4}$ $c = 32\frac{1}{4}$ Find b.   _____

7. The antenna pole shown is 60 feet high, and the guy wire is attached 5 feet below the top of the pole. What is the length of the guy wire if it is to be fastened 22 feet from the base of the antenna pole?   _____

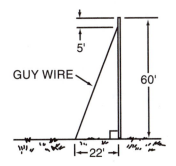

8. The 24-foot ladder is used to make a service entrance connection on a building. The bottom of the ladder cannot be more than 6 feet from the building. What is the minimum height at which the ladder will be touching the building?   _____

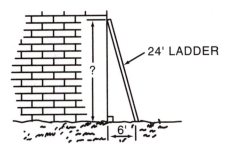

9. The 15-mile power line shown can be straightened out with a new right of way. How long will the new line be? _____

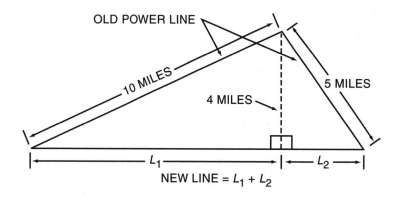

OLD POWER LINE

10 MILES

4 MILES

5 MILES

L_1

L_2

NEW LINE = $L_1 + L_2$

UNIT 41   TRIGONOMETRIC FUNCTIONS

BASIC PRINCIPLES OF TRIGONOMETRIC FUNCTIONS

A *triangle* is a closed, three-sided figure. Every triangle, no matter what its shape, has six parts: three sides and three angles. The triangle shown in Figure 1 is a *right* triangle, so named because one of its angles (angle C or ∠C) is a right angle or a 90° angle. The right angle is indicated by a small box in a diagram of a right triangle. The horizontal line *b* is the *base,* and the vertical line *a* meeting the base is the *altitude*. The sloping line *c* joining the altitude and base is the *hypotenuse.*

Figure 1

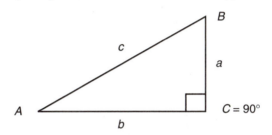

Trigonometry is the study of the measures of the sides and angles of triangles. There are three angles in a triangle. The sum of these angles will always be 180°. In the case of the right triangle, since one of the angles is 90°, the sum of the two remaining angles must also equal 90°.

The angles of a right triangle are determined by the length of its sides. These sides are called the hypotenuse, the opposite side, and the adjacent side. The hypotenuse is always the longest side and is always located opposite the right angle. The side that is opposite or adjacent to the other two angles is determined by the specific angle it is referenced to. The opposite side can be found by bisecting the particular angle of reference. If the bisect line is extended, it will intersect the opposite side. See Figure 2 for the relationships described.

Figure 2

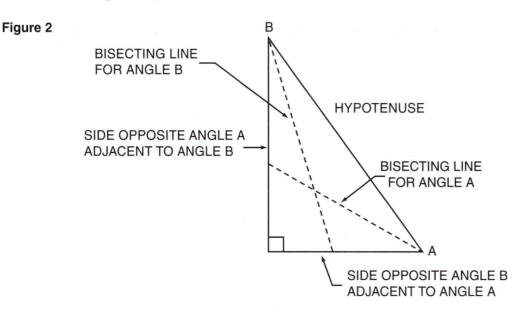

BISECTING LINE
FOR ANGLE B

B

HYPOTENUSE

SIDE OPPOSITE ANGLE A
ADJACENT TO ANGLE B →

BISECTING LINE
FOR ANGLE A

SIDE OPPOSITE ANGLE B
ADJACENT TO ANGLE A

A

The two angles that are less than 90° are formed by the hypotenuse and a side.

There is a special relationship among the sides and angles of a right triangle. Knowing any two sides of a right triangle enables us to calculate the two unknown angles (we know the third is 90°). This is accomplished by determining different ratios of the sides of the triangle with respect to an angle. These ratios are called the *sine, cosine*, and *tangent* of an angle. Table 1 on page 153 presents the values of the sine (sin), cosine (cos), and tangent (tan) for angles ranging from 1° to 90°. Most calculators are able to compute the trigonometric functions for any angle.

Sine of an Angle

The sine of an angle is the ratio of the angle's opposite side to the hypotenuse. This ratio is a constant for a given angle regardless of the size of the right triangle. Referring to Figure 1, the sine of ∠A is:

$$\sin A = \frac{opp}{hyp} = \frac{a}{c}$$

Example: A right triangle has an altitude of 15 inches and a hypotenuse of 22 inches. Locate the angle in Table 1 that is the closest to the value calculated.

$$\sin A = \frac{opp}{hyp} = \frac{15\ in}{22\ in} = 0.6818$$

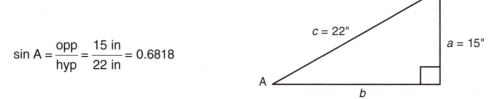

$c = 22"$

$a = 15"$

A

b

Moving down the sin column in Table 1, we find that angle A is approximately 43° as its sin = 0.6820.

Cosine of an Angle

The cosine of an angle is the ratio of the angle's adjacent side to the hypotenuse. This ratio is a constant for a given angle regardless of the size of the right triangle. Referring to Figure 1, the cosine of ∠A is:

$$\cos A = \frac{adj}{hyp} = \frac{b}{c}$$

Example: A right triangle has a base of 12 inches and a hypotenuse of 18 inches. Locate the angle in Table 1 that is the closest to the value calculated.

$$\sin A = \frac{adj}{hyp} = \frac{12\ in}{18\ in} = 0.6667$$

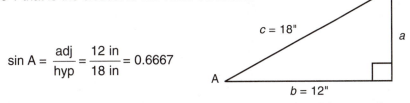

Moving down the cos column in Table 1, we find that angle A is approximately 48° as its cos = 0.6691.

Tangent of an Angle

The tangent of an angle is the ratio of the angle's opposite side to its adjacent side. This ratio is a constant for a given angle regardless of the size of the right triangle. Referring to Figure 1, the tangent of ∠A is:

$$\tan A = \frac{opp}{adj} = \frac{a}{b}$$

Example: A right triangle has a base of 20 centimeters and a altitude of 11.55 centimeters. Locate the angle in Table 1 which is the closest to the value calculated.

$$\tan A = \frac{opp}{adj} = \frac{11.55\ cm}{20\ cm} = 0.5775$$

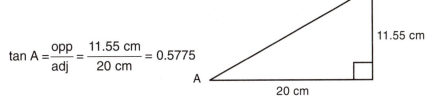

Moving down the tan column in Table 1, we find that angle A is approximately 30° as its tan = 0.5774.

Table 1

Angle	sin	cos	tan	Angle	sin	cos	tan
1°	0.017 5	0.999 8	0.017 5	46°	0.719 3	0.694 7	1.035 5
2°	0.034 9	0.999 4	0.034 9	47°	0.731 4	0.682 0	1.072 4
3°	0.052 3	0.998 6	0.052 4	48°	0.743 1	0.669 1	1.110 6
4°	0.069 8	0.997 6	0.069 9	49°	0.754 7	0.656 1	1.150 4
5°	0.087 2	0.996 2	0.087 5	50°	0.766 0	0.642 8	1.191 8
6°	0.104 5	0.994 5	0.105 1	51°	0.777 1	0.629 3	1.234 9
7°	0.121 9	0.992 5	0.122 8	52°	0.788 0	0.615 7	1.279 9
8°	0.139 2	0.990 3	0.140 5	53°	0.798 6	0.601 8	1.327 0
9°	0.156 4	0.987 7	0.158 4	54°	0.809 0	0.587 8	1.376 4
10°	0.173 6	0.984 8	0.176 3	55°	0.819 2	0.573 6	1.428 1
11°	0.190 8	0.981 6	0.194 4	56°	0.829 0	0.559 2	1.482 6
12°	0.207 9	0.978 1	0.212 6	57°	0.838 7	0.544 6	1.539 9
13°	0.225 0	0.974 4	0.230 9	58°	0.848 0	0.529 9	1.600 3
14°	0.241 9	0.970 3	0.249 3	59°	0.857 2	0.515 0	1.664 3
15°	0.258 8	0.965 9	0.267 9	60°	0.866 0	0.500 0	1.732 1
16°	0.275 6	0.961 3	0.286 7	61°	0.874 6	0.484 8	1.804 0
17°	0.292 4	0.956 3	0.305 7	62°	0.882 9	0.469 5	1.880 7
18°	0.309 0	0.951 1	0.324 9	63°	0.891 0	0.454 0	1.962 6
19°	0.325 6	0.945 5	0.344 3	64°	0.898 8	0.438 4	2.050 3
20°	0.342 0	0.939 7	0.364 0	65°	0.906 3	0.422 6	2.144 5
21°	0.358 4	0.933 6	0.383 9	66°	0.913 5	0.406 7	2.246 0
22°	0.374 6	0.927 2	0.404 0	67°	0.920 5	0.390 7	2.355 9
23°	0.390 7	0.920 5	0.424 5	68°	0.927 2	0.374 6	2.475 1
24°	0.406 7	0.913 5	0.445 2	69°	0.933 6	0.358 4	2.605 1
25°	0.422 6	0.906 3	0.466 3	70°	0.939 7	0.342 0	2.747 5
26°	0.438 4	0.898 8	0.487 7	71°	0.945 5	0.325 6	2.904 2
27°	0.454 0	0.891 0	0.509 5	72°	0.951 1	0.309 0	3.077 7
28°	0.469 5	0.882 9	0.531 7	73°	0.956 3	0.292 4	3.270 9
29°	0.484 8	0.874 6	0.554 3	74°	0.961 3	0.275 6	3.487 4
30°	0.500 0	0.866 0	0.577 4	75°	0.965 9	0.258 8	3.732 1
31°	0.515 0	0.857 2	0.600 9	76°	0.970 3	0.241 9	4.010 8
32°	0.529 9	0.848 0	0.624 9	77°	0.974 4	0.225 0	4.331 5
33°	0.544 6	0.838 7	0.649 4	78°	0.978 1	0.207 9	4.704 6
34°	0.559 2	0.829 0	0.674 5	79°	0.981 6	0.190 8	5.144 6
35°	0.573 6	0.819 2	0.700 2	80°	0.984 8	0.173 6	5.671 3
36°	0.587 8	0.809 0	0.726 5	81°	0.987 7	0.156 4	6.313 8
37°	0.601 8	0.798 6	0.753 6	82°	0.990 3	0.139 2	7.115 4
38°	0.615 7	0.788 0	0.781 3	83°	0.992 5	0.121 9	8.144 3
39°	0.629 3	0.777 1	0.809 8	84°	0.994 5	0.104 5	9.514 4
40°	0.642 8	0.766 0	0.839 1	85°	0.996 2	0.087 2	11.430 1
41°	0.656 1	0.754 7	0.869 3	86°	0.997 6	0.069 8	14.300 7
42°	0.669 1	0.743 1	0.900 4	87°	0.998 6	0.052 3	19.081 1
43°	0.682 0	0.731 4	0.932 5	88°	0.999 4	0.034 9	28.636 3
44°	0.694 7	0.719 3	0.965 7	89°	0.999 8	0.017 5	57.290 0
45°	0.707 1	0.707 1	1.000 0	90°	1.000 0	0.000 0	

FINDING TRIGONOMETRIC FUNCTIONS WITH A CALCULATOR

Scientific calculators are equipped with trigonometric function keys: sine **(SIN)**, cosine **(COS)**, and tangent **(TAN)**. These keys are used to find the sine, cosine, and tangent of different angles. When one of these keys is pressed, it will give the SIN, COS, or TAN of any angle shown on the *x* axis or display.*
Note: Make certain that the calculator is set for degrees, not gradians or radians.

Example: Enter 55 on the display and press the **SIN** key.

The number 0.819152044 (depending on the number of digits your calculator can display) will be seen. This is the sine of a 55° angle.

Clear the display and press 55 again. This time press the **COS** key.

This time 0.573576436 is shown on the display. This is the cosine of a 55° angle.

Finding the Angle

The trig function keys can also be used in a reverse operation. They can be used to find an angle when the sine, cosine, or tangent of the angle is known. Different brands of calculators accomplish this in different ways. Some use a second function key **(2nd)** and display the operation as follows:

$$\text{SIN}^{-1} \quad \text{COS-1} \quad \text{TAN}^{-1}$$

Other calculators have an invert **(INV)** or **ARC** key. All of these perform the same function. Assume that it is known that the cosine of an angle is 0.4556, and it is necessary to find the angle that corresponds to this cosine.

Enter .4556

| 0.4556 |

Press 2nd, INV, or ARC (depending on the calculator)

Press COS

| 62.89645407 |

The number 62.89645407 will be displayed. This is the degree angle that corresponds to a cosine of 0.4556. If the 2nd, INV, or ARC key is not pressed before COS, the number 0.999968385 will be displayed. This is the cosine of a 0.4556° angle.

*On more complex calculators, such as graphing calculators, the trig function keys are pressed before the angle or ratio is entered.

A simple memory aid that can be used to help remember the relationship of the sides of a right triangle to the sine, cosine, and tangent of an angle is: **Oscar Had A Heap Of Apples.** To use this memory aid, write down sin, cos, and tan. The first letter of each word in the saying gives the relationship of the sides to the trigonometric function.

$$\textbf{Sin} \quad \frac{\textbf{Oscar}}{\textbf{Had}} \quad \frac{\text{(opposite)}}{\text{(hypotenuse)}}$$

$$\textbf{Cos} \quad \frac{\textbf{A}}{\textbf{Heap}} \quad \frac{\text{(adjacent)}}{\text{(hypotenuse)}}$$

$$\textbf{Tan} \quad \frac{\textbf{Of}}{\textbf{Apples}} \quad \frac{\text{(opposite)}}{\text{(adjacent)}}$$

PRACTICAL PROBLEMS

Round the answer to the nearer tenth.

Note: Use this illustration for problems 1–6.

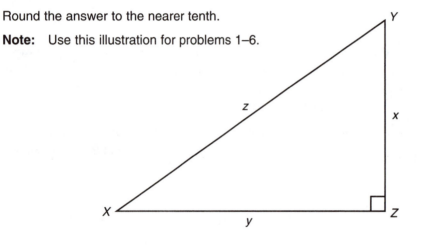

1. $z = 10$ feet $y = 7$ feet   Find $\angle X$.   _____

2. $z = 47$ feet $x = 23$ feet   Find $\angle X$.   _____

3. $y = 5.42$ feet $x = 3.3$ feet   Find $\angle X$.   _____

4. $y = 13.5$ inches $x = 21.6$ inches   Find $\angle X$.   _____

5. $x = 15$ feet $\angle Y = 27°$ Find z.   _____

6. $y = 33$ feet $z = 34.7$ feet   Find $\angle Y$.   _____

7. A roof on a storage shed is 7.3 feet wide with an incline of 20° for drainage.
    What is the height *(h)* of the roof?   _____

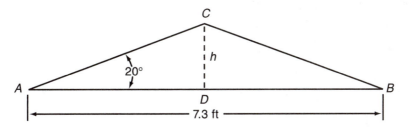

8. A loading ramp 8 feet long is placed from ground level to the top of a loading
    platform. The platform is 1.5 feet high. What is the angle *(A)* of incline?   _____

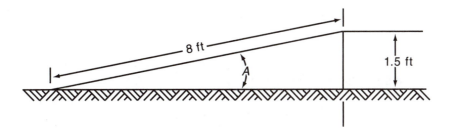

9. An electrician is to bend a pipe to make a 2.2-foot rise in a 5.85-foot distance.   _____

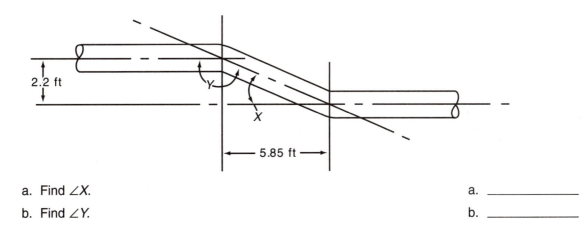

     a. Find ∠*X.*   a. _____

     b. Find ∠*Y.*   b. _____

10. A furnace 6 feet wide and 4.5 feet high is placed in an attic with the front side vertically in line with the roof apex. The gabled roof has an incline of 22 degrees and covers a width span of 36 feet. What is the clearance *d* between the top back edge of the furnace and the roof?

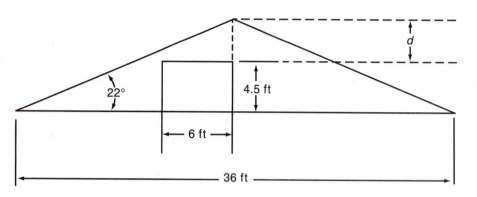

11. A 1-foot ruler held perpendicular to a flat surface casts a shadow 8 inches long. At the same time, a pole casts a shadow 18 feet 7 inches long. What is the height of the pole?

UNIT 42   PLANE VECTORS

BASIC PRINCIPLES OF PLANE VECTORS

A *vector* graphically represents any quantity that has both *magnitude* and *direction.* Voltages and currents of electricity are directed quantities and can be expressed as vectors. Two or more vectors may be used to represent quantities present at the same instant of time. The *resultant* is a single vector that is equivalent to the sum or difference of these vectors.

- The resultant of quantities having the same direction or phase relationship is found by adding the quantities of the first vector (V_1) and the second vector (V_2).

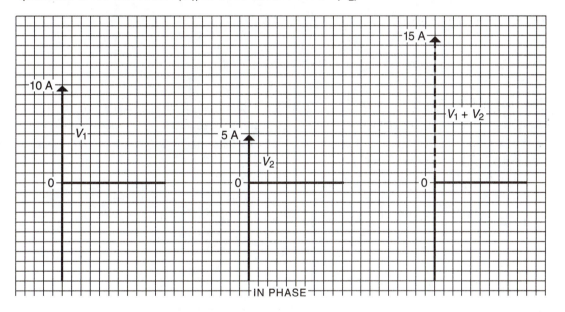

- The resultant of quantities having opposite direction, or being 180° out of phase, is found by subtracting the quantities, $V_1 - V_2$.

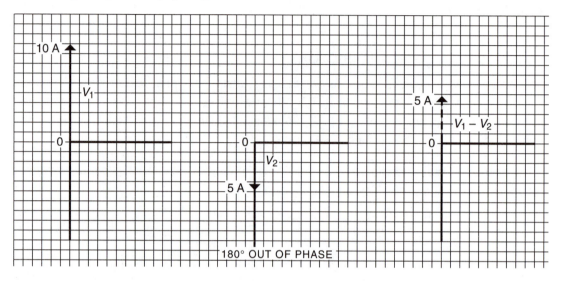

- The resultant of quantities that are not parallel is found to be the quantity represented by the diagonal of the parallelogram whose two adjacent sides represent the quantities both in magnitude and direction. The Pythagorean theorem is used to find the magnitude of the resultant vector when V_1 and V_2 are 90° out of phase.

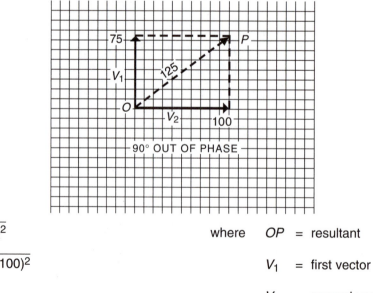

$$OP = \sqrt{V_1^2 + V_2^2}$$

$$= \sqrt{(75)^2 + (100)^2}$$

$$= 125$$

where OP = resultant

V_1 = first vector

V_2 = second vector

- The law of cosines is used to find the magnitude of the resultant when V_1 and V_2 are less than 90° out of phase or greater than 90° out of phase. The two versions of the law of cosines for the two situations follow.

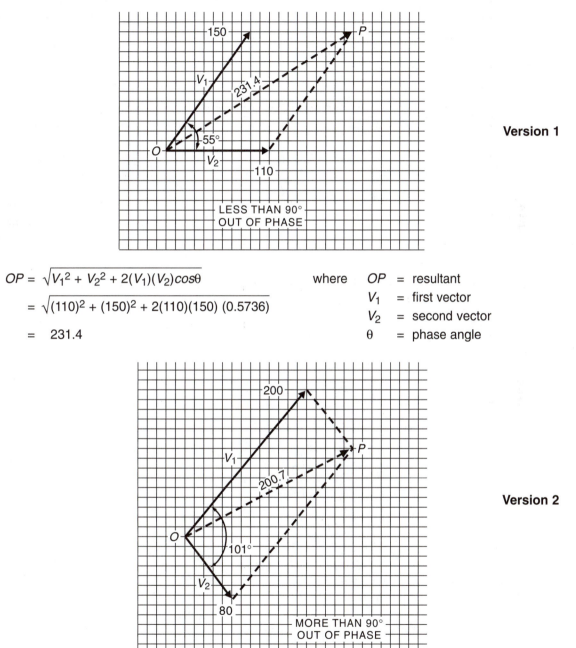

Version 1

$$OP = \sqrt{V_1{}^2 + V_2{}^2 + 2(V_1)(V_2)\cos\theta}$$

$$= \sqrt{(110)^2 + (150)^2 + 2(110)(150)\,(0.5736)}$$

$$= 231.4$$

where
OP = resultant
V_1 = first vector
V_2 = second vector
θ = phase angle

Version 2

$$OP = \sqrt{V_1{}^2 + V_2{}^2 - 2(V_1)(V_2)\cos(180° - \theta)}$$

$$= \sqrt{(80)^2 + (200)^2 - 2(80)(200)(0.1908)}$$

$$= 200.7$$

where		
OP	=	resultant
V_1	=	first vector
V_2	=	second vector
θ	=	phase angle

Study the electrical impedance diagram.

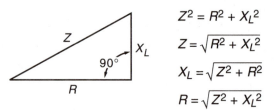

$$Z^2 = R^2 + X_L{}^2$$

$$Z = \sqrt{R^2 + X_L{}^2}$$

$$X_L = \sqrt{Z^2 + R^2}$$

$$R = \sqrt{Z^2 + X_L{}^2}$$

where		
Z	=	impedance
X_L	=	inductive reactance
R	=	resistance

Study the voltage diagram.

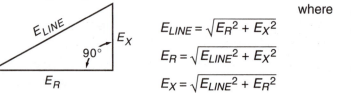

$$E_{LINE} = \sqrt{E_R{}^2 + E_X{}^2}$$

$$E_R = \sqrt{E_{LINE}{}^2 + E_X{}^2}$$

$$E_X = \sqrt{E_{LINE}{}^2 + E_R{}^2}$$

where		
E_{LINE}	=	line voltage
E_R	=	voltage across the resistance
E_X	=	voltage of the reactance

PRACTICAL PROBLEMS

Express the answer to the nearer hundredth when necessary.

Note: Use this illustration for problems 1–3.

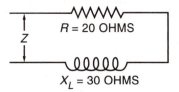

R = 20 OHMS

X_L = 30 OHMS

1. Find the impedance *(Z)* of the circuit shown if the inductive reactance *(X_L)* is 30 ohms and the resistance *(R)* is 20 ohms. _____

2. If the resistance is changed to 40 ohms, what is the impedance? _____

3. The impedance *(Z)* is 20 ohms and the resistance *(R)* is 10 ohms. What is the inductive reactance *(X$_L$)*? _____

4. What is the value of the resistance *(R)* in the series circuit shown? _____

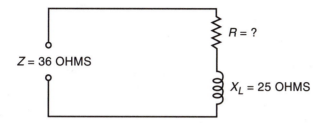

Note: Use this diagram for problems 5–7.

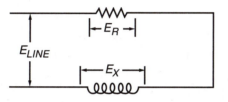

5. Resistance voltage *(E$_R$)* in the voltage diagram is 14 volts, and reactance voltage *(E$_X$)* is 18 volts. What is the line voltage *(E$_{LINE}$)*? _____

6. The line voltage *(E$_{LINE}$)* is 120 volts, and the voltage of the reactance *(E$_X$)* is 60 volts. What is the voltage across the resistance *(E$_R$)*? _____

7. Reactance voltage *(E$_X$)* is 150 volts, and resistance voltage *(E$_R$)* is 200 volts. Find the voltage across the line *(E$_{LINE}$)*. _____

8. In the circuit shown, *E$_c$* is 135 volts and *E$_b$* is 175 volts. Find *E$_a$*. _____

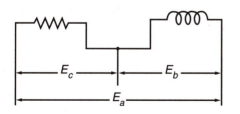

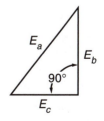

VOLTAGE DIAGRAM

Note: Use this illustration for problems 9–10.

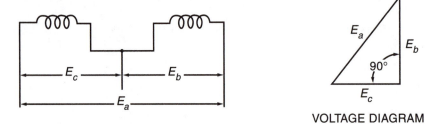

VOLTAGE DIAGRAM

9. This diagram represents a 3-wire, 2-phase voltage source. If E_b is 130 volts and E_a is 182 volts, find E_c.

    _____

10. Determine the voltage of E_a if E_b is 105 volts and E_c is 120 volts.

    _____

11. A motor draws 3.8 amperes of current when operated from a 120-volt, 60-hertz power source. The motor winding has a series resistance (R) of 30 ohms and an inductive reactance (X_L) of 10 ohms. Find the line voltage to the nearer whole volt. (*Hint:* First apply Ohm's law $E = IR$ to find E_R and E_X.)

    _____

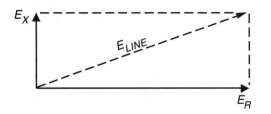

12. A transformer secondary circuit is shown. Find I_s. Express the answer to the nearer tenth. (*Hint:* First apply Ohm's law in form $I = E/R$ to find I_R and I_L.)

    _____

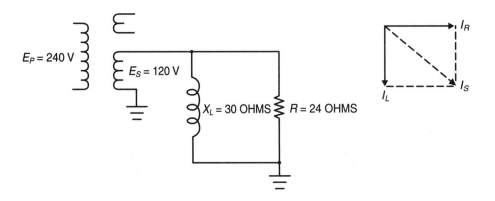

13. A 20-ohm resistor and a capacitor of 15 ohms capacitive reactance at 60 hertz are connected in parallel across a 60-volt, 60-hertz source. Find I_{LINE} to the nearer tenth. (*Hint*: First apply Ohm's Law.)

_____

14. Two voltages (E_1 and E_2) are applied to a resistor. E_1 is 40 volts at 0° reference. E_2 is 60 volts and leads E_1 by 40°. Find E_R to the nearer tenth.

_____

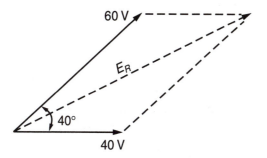

15. Two voltages (E_1 and E_2) are applied to a resistor. E_1 is 60 volts at 0° reference and E_2 is 90 volts and leads E_1 by 104°. Find E_R to the nearer tenth.

_____

UNIT 43   ROTATING VECTORS

BASIC PRINCIPLES OF ROTATING VECTORS

The concept of vectors rotating in a circular motion is used in the study of alternating current and voltage. The vector is pictured as the radius of a circle and is rotated in a counterclockwise direction at a constant speed. When the vector is rotated from A to B, the angle between the reference axis and the vector position at B is 45°. At position C the angle is 90°.

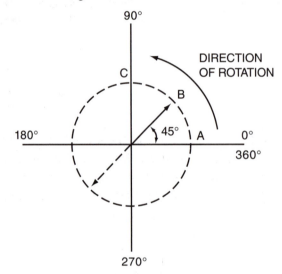

- The rotating vector may be resolved into its vertical component for each degree of rotation or any instant in time. The magnitude of the vertical component is plotted as a sine curve.

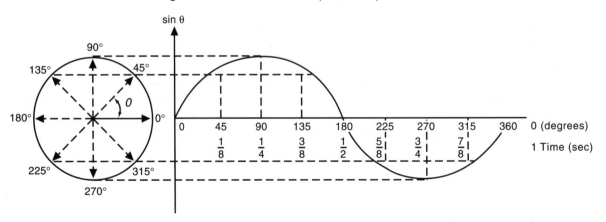

Note: On the diagram, E_{max} corresponds to the highest point on the sine curve.

- An angular rotation of 360° is one *cycle*. The *frequency* is the number of complete cycles per second. In the diagram above, the frequency is shown to be 1 cycle/sec (1 hertz). Notice that angle measure and time are marked on the same axis.

Study the formula for finding frequency.

$$f = \frac{1}{t}$$

where f = frequency in hertz
t = time in seconds for one rotation or one cycle

Study the formula for finding instantaneous voltage.

$$E_i = E_{max}\sin\theta$$

where E_i = instantaneous voltage
E_{max} = maximum voltage
θ = angle of rotation

Study the formula for finding instantaneous current.

$$I_i = E_{max}\sin\theta$$

where I_i = instantaneous voltage
I_{max} = maximum current
θ = angle of rotation

The voltage curves of two AC generators 90° out of phase are illustrated. The rotating vectors to the left of the voltage curves represent the AC generators. Generator 2 is started one-fourth of a cycle after generator 1. Their generated voltages are separated by the period of time corresponding to 90°. This is the *phase angle* (θ) between the two voltage generations. Notice that in this diagram, the phase angle corresponds to t_i = 4 milliseconds.

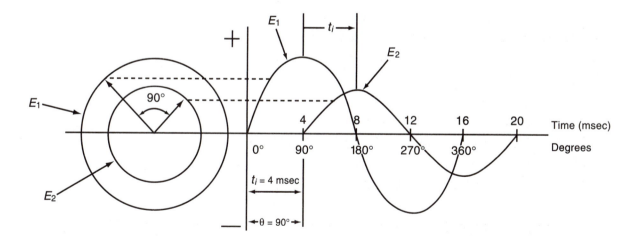

Study the formula for finding the phase angle.

$$\theta = 360t_i f$$

where
θ	=	phase angle in degrees between the two voltages
t_i	=	time interval in seconds between the two voltages
f	=	frequency of the AC generators
360	=	number of degrees in one cycle

PRACTICAL PROBLEMS

1. An AC current has a peak value of 50 amperes. Find the instantaneous value of the current at 10 degrees. Express the answer to the nearer hundredth. _____

2. What is the magnitude of voltage at point A? Express the answer to the nearer tenth. _____

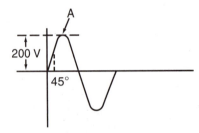

3. An AC voltage wave has an instantaneous value of 80 volts at 30 degrees. Find the maximum voltage value of the wave. _____

4. An AC voltage has an instantaneous value of 110 volts. Find to the nearer degree the phase angle if the peak voltage is 155.5 volts. _____

5. An AC current has a value of 18 amperes at 12 degrees. What is the value at 90 degrees? Round the answer to the nearer hundredth. _____

6. Find the phase angle at which an instantaneous voltage of 144 volts appears in a wave with peak value of 500 volts. Round the answer to the nearer degree. _____

7. The voltage of an AC wave is 100 volts at 30 degrees. Find the instantaneous voltage at each value.

    a. 0°   a. _____

    b. 15°   b. _____

    c. 60°   c. _____

    d. 75°   d. _____

    e. 90°   e. _____

8. What is the time in milliseconds required to generate one cycle of voltage at 60 hertz?   _____

9. In 10 microseconds, a generator produces one hertz of voltage. What is the frequency of the generator?   _____

10. What is the phase angle between two 60-hertz voltage generations separated by a time interval of three milliseconds?   _____

UNIT 44   COMBINED PROBLEMS IN TRIGONOMETRY

This unit provides practical problems in trigonometry. Use the formulas provided in previous units to solve these problems.

PRACTICAL PROBLEMS

Solve problems 1–5 for the remaining part of the right triangle shown in the figure. Round the answer to the nearer hundredth.

1.   If $a = 4$, $b = 8$,   find c.   _____

2.   If $b = 25$, $c = 40$,   find a.   _____

3.   If $c = 350$, $a = 200$,   find b.   _____

4.   If $a = 7.5$, $b = 12.5$,   find c.   _____

5.   If $b = 12.4$, $c = 67.8$,   find a.   _____

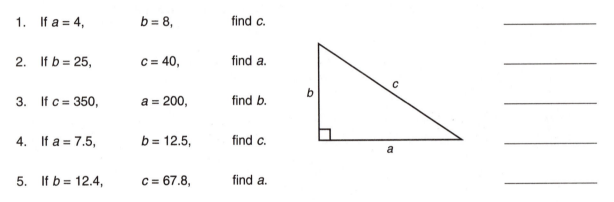

Note: Using the impedance diagram, solve for the missing quantities in each series circuit for problems 6–10. Round the answer to the nearer tenth.

Z = impedance

R = resistance

X_L = inductive reactance

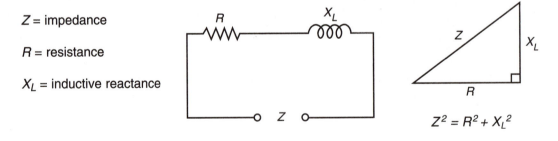

$$Z^2 = R^2 + X_L^2$$

6. $X_L = 500$ ohms $R = 500$ ohms   _____

7. $R = 45$ ohms $Z = 90$ ohms   _____

8. $Z = 1{,}100$ ohms $X_L = 400$ ohms   _____

9. $X_L = 7.2$ ohms $R = 10.2$ ohms   _____

10. In the series circuit, if the impedance is 70 ohms and the reactance is 50 ohms, what is the value of the resistance? Round the answer to the nearer ohm.

_____

Note: Use this diagram for problems 11–12. Round the answers to the nearer tenth.

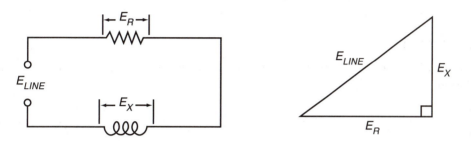

11. What is the voltage across the reactance (E_X) if $E_{LINE} = 240$ volts and $E_R = 120$ volts?

_____

12. If the voltage across the reactance (E_X) is 50 volts and the line voltage (E_{LINE}) is 300 volts, what is the voltage across the resistance (E_R)?

_____

13. What is the resultant of two vectors if $OA = 7$ amperes, $OB = 10$ amperes, and the phase angle is 50°? Round the answer to the nearer tenth.

_____

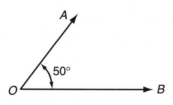

14. What is the output frequency to the nearer tenth hertz of a generator that produces two voltage waveforms in 5.55 milliseconds?

_____

15. What is the phase difference in degrees between the voltages across R_L if generator 1 is started 2.77 milliseconds before generator 2? The frequency of both generators is 120 hertz. Round the answer to the nearer degree.

    _____

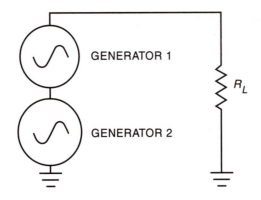

16. What is the maximum peak value of a 60-hertz wave that has an instantaneous voltage of 22 volts 1.67 milliseconds before the end of one cycle? Round the answer to the nearer tenth.

    _____

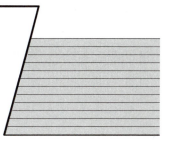

Appendix

SECTION I

DENOMINATE NUMBERS

Denominate numbers are numbers that include units of measurement. The units of measurement are arranged from the largest unit at the left to the smallest unit at the right.

Example: 6 yd 2 ft 4 in is the correct representation of a denominate number.

All basic operations of arithmetic can be performed on denominate numbers.

I. EQUIVALENT MEASURES

A measurement can be expressed in an equivalent form, but with different units, through multiplication by a *conversion factor.* For example, the conversion factor that changes feet to inches, or inches to feet, is

$$12 \text{ in} = 1 \text{ ft}$$

To change a measurement given in inches to one in feet, multiply by the conversion factor $\dfrac{1 \text{ ft}}{12 \text{ in}}$.

To change a measurement given in feet to one in inches, multiply by the conversion factor $\dfrac{12 \text{ in}}{1 \text{ ft}}$.

Example: To express 6 inches in equivalent foot measurement, multiply by $\dfrac{1 \text{ ft}}{12 \text{ in}}$. In the numerator and demnominator, divide by a common factor.

$$6 \text{ in} = \frac{\overset{1}{\cancel{6 \text{ in}}}}{1} \times \frac{1 \text{ ft}}{\underset{2}{\cancel{12 \text{ in}}}} = \frac{1}{2} \text{ ft or } 0.5 \text{ ft}$$

To express 4 feet in equivalent inch measurement, multiply 4 feet by $\dfrac{12 \text{ in}}{1 \text{ ft}}$.

$$4 \text{ ft} = \frac{\overset{4}{\cancel{4 \text{ ft}}}}{1} \times \frac{12 \text{ in}}{\underset{1}{\cancel{1 \text{ ft}}}} \quad \frac{48 \text{ in}}{1} = 48 \text{ in}$$

Per means division, as with a fraction bar. For example, 50 miles per hour can be written $\dfrac{50 \text{ miles}}{1 \text{ hour}}$.

II.   BASIC OPERATIONS

A.   Addition

Example: 2 yd 1 ft 5 in + 1 ft 8 in + 5 yd 2 ft

1. Write the denominate numbers in a column with like units in the same column.

$$
\begin{array}{rlll}
 & 2 \text{ yd} & 1 \text{ ft} & 5 \text{ in} \\
 & & 1 \text{ ft} & 8 \text{ in} \\
+ & 5 \text{ yd} & 2 \text{ ft} & \\
\hline
\end{array}
$$

2. Add the denominate numbers in each column.

$$
\begin{array}{rlll}
 & 7 \text{ yd} & 4 \text{ ft} & 13 \text{ in}
\end{array}
$$

3. Express the answer using the largest possible units.

7 yd			=	7 yd		
	4 ft		=	1 yd	1 ft	
		13 in	= +		1 ft	1 in
7 yd	4 ft	13 in	=	8 yd	2 ft	1 in

B.   Subtraction

Example: 4 yd 3 ft 5 in − 2 yd 1 ft 7 in

1. Write the denominate numbers in columns with like units in the same column.

$$
\begin{array}{rlll}
 & 4 \text{ yd} & 3 \text{ ft} & 5 \text{ in} \\
- & 2 \text{ yd} & 1 \text{ ft} & 7 \text{ in} \\
\hline
\end{array}
$$

2. Starting at the right, examine each column to compare the numbers. If the bottom number is larger, exchange one unit from the column at the left for its equivalent. Combine like units.

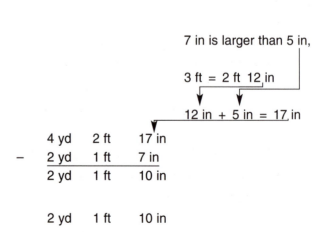

7 in is larger than 5 in,

3 ft = 2 ft 12 in

12 in + 5 in = 17 in

3. Subtract the denominate numbers.

$$
\begin{array}{rlll}
 & 4 \text{ yd} & 2 \text{ ft} & 17 \text{ in} \\
- & 2 \text{ yd} & 1 \text{ ft} & 7 \text{ in} \\
\hline
 & 2 \text{ yd} & 1 \text{ ft} & 10 \text{ in}
\end{array}
$$

4. Express the answer using the largest possible units.

$$
\begin{array}{rlll}
 & 2 \text{ yd} & 1 \text{ ft} & 10 \text{ in}
\end{array}
$$

C. Multiplication

—By a constant

Example: 1 hr 24 min × 3

1. Multiply each denominate number
    by the constant.

$$\begin{array}{cc} \text{1 hr} & \text{24 min} \\ & \times\ 3 \\ \hline \text{3 hr} & \text{72 min} \end{array}$$

2. Express the answer using the
    largest possible units.

3 hr		=	3 hr	
	72 min	=	1 hr	12 min
3 hr	72 min	=	4 hr	12 min

—By a denominate number expressing linear measurement

Example: 9 ft 6 in × 10 ft

1. Express all denominate numbers
    in the same unit.

$$9 \text{ ft } 6 \text{ in} = 9\frac{1}{2}\text{ ft}$$

2. Multiply the denominate numbers.
    (This includes the units of measure,
    such as ft × ft = sq ft.)

$$9\frac{1}{2}\text{ ft} \times 10 \text{ ft} =$$

$$\frac{19}{2}\text{ ft} \times 10 \text{ ft} = 95 \text{ sq ft}$$

—By a denominate number expressing square measurement

Example: 3 ft × 6 sq ft

1. Multiply the denominate numbers.
    (This includes the units of measure,
    such as ft × ft = sq ft and sq ft × ft = cu ft.)

$$3 \text{ ft} \times 6 \text{ sq ft} = 18 \text{ cu ft}$$

—By a denominate number expressing rate

Example: 50 mi per hr × 3 hr

1. Express the rate as a fraction using
    the fraction bar for *per*.

$$\frac{50 \text{ mi}}{1 \text{ hr}} \times \frac{3 \text{ hr}}{1} =$$

2. Divide the numerator and denominator
    by any common factors, including units
    of measure.

$$\frac{50 \text{ mi}}{1 \text{ hr}} \times \frac{\overset{3}{\cancel{3 \text{ hr}}}}{1} =$$

3. Multiply numerators.
    Multiply denominators.

$$\frac{150 \text{ mi}}{1} =$$

4. Express the answer in the remaining unit.

150 mi

D. Division

—By a constant

Example: 8 gal 3 qt ÷ 5

1. Express all denominate numbers
    in the same unit.

8 gal 3 qt = 35 qt

2. Divide the denominate number
    by the constant.

35 qt ÷ 5 = 7 qt

3. Express the answer using the
    largest possible units.

7 qt = 1 gal 3 qt

—By a denominate number expressing linear measurement

Example: 11 ft 4 in ÷ 8 in

1. Express all denominate numbers
    in the same unit.

11 ft 4 in = 136 in

2. Divide the denominate numbers
    by a common factor. (This includes
    the units of measure, such as
    inches ÷ inches = 1.)

136 in ÷ 8 in =

$$\frac{\overset{17}{\cancel{136 \text{ in}}}}{\underset{1}{\cancel{8 \text{ in}}}} = \frac{17}{1} = 17$$

—By a linear measure with a square measurement as the dividend

Example: 20 sq ft ÷ 4 ft

1. Divide the denominate numbers. (This includes the units of measure, such as sq ft ÷ ft = ft.)

   20 sq ft ÷ 4 ft

   5 ft

   $$\frac{20\ \cancel{sq\ ft}}{\cancel{4\ ft}} = \frac{5\ ft}{1}$$

   1

2. Express the answer in the remaining unit.

   5 ft

—By denominate numbers used to find rate

Example: 200 mi ÷ 10 gal

1. Divide the denominate numbers.

   20 mi

   $$\frac{\cancel{200\ mi}}{10\ gal} = \frac{20\ mi}{1\ gal}$$

   $\cancel{1\ gal}$

2. Express the units with the fraction bar meaning *per*.

   $$\frac{20\ mi}{1\ gal} = 20\ \text{miles per gallon}$$

Note: Alternative methods of performing the basic operations will produce the same results. The choice of method is determined by the individual.

SECTION II

EQUIVALENTS

ENGLISH RELATIONSHIPS

ENGLISH LENGTH MEASURE

1 foot (ft)	=	12 inches (in)
1 yard (yd)	=	3 feet (ft)
1 mile (mi)	=	1,760 yards (yd)
1 mile (mi)	=	5,280 feet (ft)

ENGLISH AREA MEASURE

1 square yard (sq yd)	=	9 square feet (sq ft)
1 square foot (sq ft)	=	144 square inches (sq in)
1 square mile (sq mi)	=	640 acres
1 acre	=	43,560 square feet (sq ft)

ENGLISH VOLUME MEASURE FOR SOLIDS

1 cubic yard (sq yd)	=	27 cubic feet (cu ft)
1 cubic foot (cu ft)	=	1,728 cubic inches (cu in)

ENGLISH VOLUME MEASURE FOR FLUIDS

1 quart (qt)	=	2 pints (pt)
1 gallon (gal)	=	4 quarts (qt)

ENGLISH VOLUME MEASURE EQUIVALENTS

1 gallon (gal)	=	0.133681 cubic foot (cu ft)
1 gallon (gal)	=	231 cubic inches (cu in)

SI METRICS STYLE GUIDE

SI metrics is derived from the French name *Système international d'unités.* SI metrics attempts to standardize names and usages so that students of metrics will have a universal knowledge of the application of terms, symbols, and units.

The English system of measurement (used in the United States) has always had many units in its weights and measures tables that were not applied to everyday use. For example, the pole, perch, furlong, peck, and scruple are not used often. These measurements, however, are used to form other measurements, and it has been necessary to include them in the tables. Including these measurements aids in the understanding of the sequence of measurements greater or smaller than the less frequently used units.

The metric system also has units that are not used in everyday application. Only by learning the lesser-used units is it possible to understand the order of the metric system. SI metrics, however, places an emphasis on the most frequently used units.

In using the metric system and writing its symbols, certain guidelines are followed. For the student's reference, some of the guidelines are listed.

1. In using the symbols for metric units, the first letter is capitalized only if it is
    derived from the name of a person.

Example:	UNIT	SYMBOL	UNIT	SYMBOL
	meter	m	Newton	N (named after Sir Isaac Newton)
	gram	g	degree Celsius	°C (named after Anders Celsius)

2. Prefixes usually are written with lowercase letters.

Example:	PREFIX	UNIT	SYMBOL
	centi	meter	cm
	milli	gram	mg

There are some exceptions, however.

Example:	PREFIX	UNIT	SYMBOL
	tera	meter	Tm (used to distinguish it from the metric ton, t)
	giga	meter	Gm (used to distinguish it from gram, g)
	mega	gram	Mg (used to distinguish it from milli, m)

3. Periods are not used in the symbols. Symbols for units are the same in the singular and the plural (no "s" is added to indicate a plural).

   Example: 1 mm *not* 1 mm.
    3 mm *not* 3 mms

4. When referring to a unit of measurement, symbols are not used. The symbol is used only when a number is associated with it.

   Example: The length of the room   *not*   The length of the room is expressed in m.
    is expressed in meters.   (*The length of the room is 25 m* is correct.)

5. When writing measurements that are less than one, a zero is written before the decimal point.

   Example: 0.25 m *not* .25 m

6. Digits are separated in groups of three, counting from the decimal point to the left and to the right.

   Example: 5 179 232 mm *not* 5,179,232 mm   0.566 23 mg *not* 0.56623 mg
    1 346.098 7 l *not* 1,346.0987 l

   A space is also left between the digits and the unit of measure.

   Example: 5 179 232 mm *not* 5 279 232mm

7. Symbols for area measure and volume measure are written with exponents.

   Example: 3 cm^2 *not* 3 sq cm   3 km^3 *not* 4 cu km

8. Metric words with prefixes are accented on the first syllable. In particular, kilometer is pronounced "kill'o-meter." This avoids confusion with words for measuring devices, which are generally accented on the second syllable, such as thermometer (ther-mom'-e-ter).

Glossary

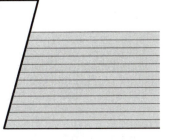

Alternating current (AC) — An electric current that reverses direction at regularly recurring intervals.

American wire gauge (AWG) — A system of numbers used in sizing wire.

Ammeter — An instrument used for measuring electric current in amperes.

Ampere (A) — A unit of electric current.

Askarel — A synthetic, noncombustible, insulating liquid used in transformers.

Ballast — A resistance used in fluorescent fixtures to stabilize the current.

Bus bar — A heavy copper strip or bar used as a primary power source to carry heavy currents or to make a common connection between several circuits.

Bushing — An insulating sleeve inserted in an opening in a metal place to protect a through conductor.

Cabinet — A flush-or surface-mounted box used as a protective housing for electrical equipment.

Cable — A stranded assembly of one or more conductors, usually within a protective sheath.

Capacitance *(C)* — The property of a conductor or capacitor that permits the storage of electrical energy. The unit of capacitance is the farad.

Capacitive reactance *(X$_c$)* — The opposition to alternating current due to capacitance. The capacitive reactance is expressed in ohms.

Capacitor — An electrical device consisting of two conducting surfaces or sets of conducting surfaces that are oppositely charged and are separated by a thin layer of insulating material such as air, paraffin paper, or mica.

Circuit — The conducting part or a system of conducting parts through which an electric current is intended to flow.

Circuit breaker — A device that automatically opens a circuit if the current exceeds the rated amount. The circuit breaker does not destroy itself as a fuse does.

Circular mil *(CM)* — The cross-sectional area of a wire that is 1 mil (0.001 inch) in diameter. Also, the unit of measure of a cross-sectional area.

Coil — An electrical element in the circuit consisting of a spiral of wire, either self-supporting or wound on a spool or other structure, usually used for electromagnetic effect or for providing resistance.

Conductance *(G)* — The ability of material to carry electrical current. It is expressed in siemens.

Conductor — A substance capable of transmitting electricity.

Conduit — A pipe or tube used for receiving and protecting electric wires and cables.

Coulomb (C) — A unit of electric charge equal to 6.25×10^{18} electrons.

Current *(I)* — The transfer of electric charge through a conductor. Current is measured in amperes.

Cycle — A complete positive and a complete negative alternation of voltage or current.

Device — An item in an electrical system carrying but not using electricity.

Dielectric — A nonconductor used between the plates of a capacitor.

Dielectric constant *(k)* — A measure of the ability of a dielectric material to store electrical potential energy.

Direct current (DC) — An electric current in which there is a continuous transfer of charge in only one direction.

Efficiency *(Eff)* — A percent value expressing the ratio of power output to power input.

Electrical metallic tubing (EMT) — A thin walled circular metal raceway used for pulling in or withdrawing wires or cables.

Electrolyte — A nonmetallic electric conductor in which current is carried by the movement of ions used in wet cell batteries.

Farad (F) — The unit of measure for capacitance.

Feeder — A heavy wire conductor connecting points of an electric distribution sytem such as a substation and a generating station.

Fitting — A small accessory, generally standardized, used for a mechanical rather than electrical function, as a locknut or bushing.

Frequency *(f)* — The number of complete cycles in a unit of time. It is expressed in hertz.

Fuse — An electrical safety device of wire or a strip of fusible metal that melts and opens the electrical circuit when the current exceeds the rated amount. It destroys itself and must be replaced.

Generator — A machine that changes mechanical energy into electrical energy of measure.

Henry (H) — The unit of measure of inductance.

Hertz (Hz) — The unit of measure for frequency.

Horsepower (hp) — A unit of power equal to 746 watts of electrical power.

Impedance *(Z)* — The total opposition to alternating current including inductive reactance, capacitive reactance, and resistance. It is expressed in ohms.

Inductance *(L)* — The property of an alternating current circuit to induce an electromotive force by variations of current. It is expressed in henrys.

Induction — The process of producing electrification, magnetization, or induced voltage in an object by exposure to a magnetic field or a charged body.

Inductive reactance *(X_L)* — The opposition to an alternating current. It is expressed in ohms.

Inductor — A conductor that acts upon another or is itself acted upon by induction. As the conductor is wound into a spiral or coil, the inductive intensity increases.

Insulator — A material that has a high resistance to the flow of an electric current.

Internal resistance — The resistance within the source of the electromotive force.

Joule (J) — A unit of electrical energy.

Junction box — A box for inserting and joining cables or wires.

Kilowatt (kW) — A unit of electrical power equal to 1,000 watts.

Kirchhoff's law — The sum of the currents entering a junction is equal to the sum of the currents leaving that junction.

Line voltage *(E_{LINE})* — The voltage existing at the wall outlets or terminals of a power line.

Live — A term that means electrically connected to a source of potential difference or electrically charged to have a potential different from that of the earth. Other terms used for live are *alive, hot,* and *energized.*

Locknut — A nut constructed to lock itself when screwed up tight.

Metric system — An international system of measurement based on powers of 10.

Mutual inductance *(M)* — The condition that exists in a circuit when the positions of two inductors cause magnetic lines of force from one inductor to link the turns of the other.

Negative (–) — A terminal or electrode with excess electrons.

Nonmetallic sheathed cable (NMC) or (NM) — A type of cable having a nonmetallic outer covering primarily in residential wiring systems.

Ohm (Ω) — A unit of measure of resistance.

Ohm's law — Current is directly proportional to voltage and inversely proportional to resistance.

Oscillator — A device for producing alternating current power at a frequency determined by the values of certain constants in the circuit.

Outlet — A set of mounted and insulated terminals to which electric appliances may be connected, such as a receptacle or an electric socket.

Parallel circuit — A method of connecting a circuit so the current has two or more paths to follow.

Positive (+) — A terminal or electrode with a deficiency of electrons.

Potential difference — The difference in electrical pressures between two points in a circuit. It is measured in volts.

Power *(P)* — The rate of doing work. It is expressed in watts.

Power factor *(PF)* — The ratio of true power (watts) to apparent power (volt amps) expressed as a percent of an AC circuit.

Primary voltage *(E_p)* — The voltage of the circuit supplying power to a transformer. It is the input voltage.

Raceway — A channel for loosely holding electrical wires.

Reactance *(X)* — Oppositon to alternating current due to inductance and capacitance. It is expressed in ohms.

Receptacle — A mounted female electrical fitting that contains the live parts of the circuit.

Resistance *(R)* — The opposition a material offers to the flow of electrons. It is expressed in ohms.

Resistor — A device that opposes the flow of an electric current. It is used for protection, operation, or current control.

Secondary voltage *(E_s)* — The voltage output of a transformer.

Series circuit — A method of connecting a circuit so the current has one path to follow.

Shunt — A conductor connected in parallel with another component in an electric circuit.

Solder — An alloy of lead and tin that melts at a low temperature and is used to join metallic surfaces.

Solenoid — An electromagnetic coil with a movable core that is drawn into the coil when the current flows.

Switch — A mechanical device for completing, interrupting, or changing the connections in an electrical circuit.

Switchboard — A type of switch-gear assembly that consists of one or more panels with mounted electrical devices.

Transformer — An electromagnetic device using induction to increase or decrease alternating current voltage.

Transistor — An electronic device consisting of a small block of semiconductors with at least three electrodes.

Volt (V) — A unit of electrical potential or pressure.

Voltage *(E)* — The electromotive force or electrical pressure. It is expressed in volts.

Voltage divider — A network of current-limiting elements connected in series to produce various voltage drops across each element.

Voltage drop — The potential difference measured across current limiting elements in a circuit. Voltage drop may be measured across a single element or several elements in a group.

Watt (W) — A unit of measure of power.

Wavelength — The distance traveled by a wave during the time interval covered by a cycle.

Wheatstone bridge — An apparatus that is used to measure resistance by varying known resistances until the system is balanced.

ANSWERS TO ODD-NUMBERED PROBLEMS

SECTION 1   WHOLE NUMBERS

UNIT 1   ADDITION OF WHOLE NUMBERS

1. 1,165	7. 2,380 ft	13. 473
3. 556	9. 615 lb	15. 1,052
5. 951	11. 14,839 W	17. 3,706 ft

UNIT 2   SUBTRACTION OF WHOLE NUMBERS

1. 1,161 ft	7. 305 ft	13. 65
3. 455	9. 200	15. 221 ft
5. 137 ft	11. 1,372 kW•h	17. 10 MΩ

UNIT 3   MULTIPLICATION OF WHOLE NUMBERS

1. a. 48	e. 42	7. 1,980 W
b. 63	f. 22	9. $19,600
c. 33	3. 16,412 W	11. 1,940 W
d. 32	5. No	13. 220 circuits

UNIT 4   DIVISION OF WHOLE NUMBERS

1. 47 staples	7. 21 ft	13. $1
3. 5,124 W	9. 8 weeks	15. 1,320 W
5. 5 outlets	11. 36 lamps	17. 30 ft

UNIT 5   COMBINED OPERATIONS WITH WHOLE NUMBERS

1. 591	11. 11,250 W	21. a. 550
3. 698 lb	13. 392	b. 75
5. 1,192 ft	15. 91	c. 400
7. 652 ft	17. 4,000 ft	d. 250
9. $238	19. 125	e. 50

SECTION 2   COMMON FRACTIONS

UNIT 6   ADDITION OF COMMON FRACTIONS

1. a. $1\frac{21}{32}$
   b. $2\frac{21}{32}$
   c. $2\frac{17}{48}$
   d. $2\frac{11}{20}$

3. $1\frac{7}{8}$ hp
5. $\frac{3}{16}$ in
7. $1\frac{7}{64}$ in
9. $\frac{7}{16}$ in

11. $6\frac{5}{16}$ in
13. $1\frac{1}{64}$ in
15. $19\frac{1}{4}$ in
17. $43\frac{9}{10}$ m

UNIT 7   SUBTRACTION OF COMMON FRACTIONS

1. 1 in
3. $\frac{25}{32}$ in
5. $\frac{9}{32}$ in

7. $2\frac{3}{8}$ in
9. $\frac{1}{64}$ in
11. $2\frac{15}{32}$ in

13. $1\frac{7}{8}$ mA

UNIT 8   MULTIPLICATION OF COMMON FRACTIONS

1. $7\frac{9}{16}$ lb
3. $255\frac{7}{12}$ ft

5. $264\frac{7}{16}$ ft
7. $26\frac{9}{16}$ kW•h

9. $\frac{5}{12}$ kg

UNIT 9   DIVISION OF COMMON FRACTIONS

1. 8
3. $1\frac{41}{49}$ W

5. 35
7. $1\frac{13}{17}$

9. $4

UNIT 10   COMBINED OPERATIONS WITH COMMON FRACTIONS

1. $2\frac{41}{48}$ hp
3. $\frac{21}{32}$ in
5. $\frac{7}{8}$ in
7. $26\frac{11}{16}$ in
9. $\frac{1}{8}$ in

11. $\frac{469}{1,000}$ in
13. $251\frac{3}{4}$ ft
15. $44\frac{1}{2}$ ft
17. $960
19. $3

21. $1,193\frac{41}{60}$ (1,193.6)
23. 30
25. 5,000 ft
27. 49

SECTION 3   DECIMAL FRACTIONS

UNIT 11   ADDITION OF DECIMAL FRACTIONS

1. $286.26	7. $122.89	13. 7 in
3. $60.05	9. 9.724 A	15. 2.6875 in
5. $82.94	11. 2.312 in	17. 4.635 cm

UNIT 12   SUBTRACTION OF DECIMAL FRACTIONS

1. 1,067.35 kW•h	9. 1.25 in	15. larger
3. 9.12 lb	11. 1.0625 in	17. #12
5. 0.6255 in	13. 1.0625 in	19. 0.0035 in
7. 3.375 in		

UNIT 13   MULTIPLICATION OF DECIMAL FRACTIONS

1. $0.171	7. 15.3615 lb	13. $2,084.58
3. 58.12 in	9. $63.35	15. 31.416 ft/min
5. 7.461 in	11. $1,194.03	17. 37.125 V

UNIT 14   DIVISION OF DECIMAL FRACTIONS

1. $1.066	7. 2.40625 in	11. $188.58
3. 0.00256 Ω	9. 180.82 W	13. 1,676.75 lb
5. $0.86		

UNIT 15   DECIMAL AND COMMON FRACTION EQUIVALENTS

1. 1.3125 in	7. 1⅝ in	13. 0.258 in
3. 25.39 lb	9. ⅛ in	15. 0.4325 in by 0.622 in
5. 0.131 in	11. 1.753 in	

UNIT 16   COMBINED OPERATIONS WITH DECIMAL FRACTIONS

1. 0.535 in	9. 3.555 in	15. $40.24
3. 2.171 A	11. 8.375 lb	17. $24.50
5. 2.2175 in	13. 1.5 A	19. 7.04 in
7. 3.86 in		

SECTION 4   PERCENTS, AVERAGES, AND ESTIMATES

UNIT 17   PERCENT AND PERCENTAGE

1.	20%	5.	592	9.	90%
3.	$112.56	7.	$94.71	11.	$161.50

UNIT 18   INTEREST

1.	$1,071	5.	$4,060.49	9.	$373.75
3.	$486	7.	$2,146.71	11.	$234

UNIT 19   DISCOUNT

1.	$203.40	5.	$74.64	9.	$1,103.42
3.	$752.06	7.	$981	11.	$1,425.60

UNIT 20   AVERAGES AND ESTIMATES

1. 51.76 kW•h

3. $177.10

5. a. $18.21
    b. $3.94 over

7. a. 8.7 coils
    b. 328.3 ft over

9. 35.7°F

UNIT 21   COMBINED PROBLEMS ON PERCENTS, AVERAGES, AND ESTIMATES

1. 16.7%

3. 82.8 hp

5. $177

7. $105

9. $265.22

11. $1,745.17

13. 94%

SECTION 5   POWERS AND ROOTS

UNIT 22   POWERS

1.	49	11.	225	21.	1,024 CM
3.	81	13.	15,625	23.	201.6 W
5.	121	15.	810,000	25.	13,444.4 W
7.	512	17.	147,008,443	27.	4,215.6877 CM
9.	10,000	19.	4,096 CM		

UNIT 23   ROOTS

1.	9	11.	32 A	21.	2.50 A
3.	13	13.	110 V	23.	101.88 mils
5.	23	15.	2 A	25.	127.28 V
7.	29.87	17.	3.95 A	27.	3.64 A
9.	41.13	19.	0.21 A	29.	110.88 mils

UNIT 24   COMBINED OPERATIONS WITH POWERS AND ROOTS

1.	11	9.	56	15.	2.77 A
3.	17	11.	10,383.61 CM	17.	3.74 A
5.	27	13.	81 mils	19.	11.52 Ω
7.	39				

UNIT 25   METRIC MEASURE AND SCIENTIFIC NOTATION

1.	one hectometer	5.	945.635045 03	9.	3,500,000 kilowatts
3.	350 μF	7.	2,450,000,000 Hz		

SECTION 6   MEASURE

UNIT 26   LENGTH MEASURE

1.	1,200 mm	9.	88.495 km	15.		3.09 m
3.	37 mm	11.	15 lengths	17.	a	8.22 cm
5.	0.43 m	13.	2.85 mm		b	13.3 cm
7.	15.24 cm					

UNIT 27   AREA MEASURE

1.	750,000 mm^2	9.	6 fixtures	17.	a.	6,000 W
3.	2.5 sq ft	11.	30.38 m^2		b.	50 A
5.	2.51 m^2	13.	61 ft		c.	4 circuits
7.	101.96 cm^2	15.	8.296 sq in			

UNIT 28   VOLUME AND MASS MEASURE

1. 1,200,000,000 mm^3
3. 1.56 cu yd
5. 5.68 L

7. 6.88 m^3
9. 0.142 m^3

11. 9,818 gal
13. 31.7 m^3

UNIT 29   ENERGY AND TEMPERATURE MEASURE

1. a. 1.5×10^3 mA
   b. 1.5×10^6 μA
3. a. 2×10^{-6} F
   b. 2×10^6 pF

5. a. 3.7×10^7 μH
   b. 3.7×10^4 mH
7. 20°C

9. 1.26 MJ
11. 8.3 msec
13. 5×10^6 m

UNIT 30   COMBINED PROBLEMS ON MEASURE

1. 304.8 mm
3. 8.045 km
5. 22.712 L
7. 11.468 m^3
9. 23.9°

11. 36 MJ
13. 0.2 MV
15. 1,245.443 L
17. 7.079 m^3

19. 0.4 cm
21. 0.75 cm
23. 25 ft
25. 0.950 mm^2

SECTION 7   RATIO AND PROPORTION

UNIT 31   RATIO

1. a.   3:1
   b.   5:2
   c.   3:1

   d.   3:1
   e.   1:3
3. 100:9

5. 1:7
7. 7:36

UNIT 32   PROPORTION

1. 61.2 min
3. 11.5 min

5. 3.192 Ω
7. 2.59 Ω

9. $74.77
11. $418.44

UNIT 33 COMBINED OPERATIONS WITH RATIO AND PROPORTION

1. $75
3. $400

5. 1:2
7. 1:20

9. 900 r/min

SECTION 8   FORMULAS

UNIT 34   REPRESENTATION IN FORMULAS

1. $R_t = R_1 + R_2 + \ldots + R_n$

3. $\dfrac{E_P}{E_S} = \dfrac{P_P}{P_S}$

5. $C = \dfrac{1}{2\pi f X_C}$

7. $P = \dfrac{E^2}{R}$

9. $L_m = k\sqrt{L_1 L_2}$

11. Conductance *(G)* is equal to the reciprocal of resistance.

13. The effective value of an AC voltage *(E)* is equal to the maximum peak value *(E_{max})* multiplied by 0.707.

15. The frequency *(f)* of an AC generator can be calculated by dividing the product of the number of pairs of poles *(P)* and the speed of the generator in revolutions per minute *(N)* by 60.

UNIT 35   REARRANGEMENT IN FORMULAS

1. $C = \dfrac{Q}{V}$

3. $Z = \sqrt{R^2 + X^2}$

5. $R_t = R_1 + R_2 + R_3$

7. $C = \dfrac{1}{2\pi f X_C}$

9. $N_P = \dfrac{N_S E_P}{E_S}$

11. $Z_P = \dfrac{Z_S N_P^2}{N_S}$

13. $r_p = \dfrac{u}{g_m}$

15. $k = \dfrac{L_m}{\sqrt{L_1 L_2}}$

UNIT 36   GENERAL SIMPLE FORMULAS

1. 40.68 Ω

3. 28.52 kW

5. 1.57 A

7. 12.5 A

9. 392° F

11. 149° F

13. 50 Ω

15. 282.5 r/min

17. 6,171 W

19. 0.926 A

21. 46.77 A

23. 12.37 Ω

25. 80 Ω

27. 83.83 kW

29. 29.92 kW

31. 22,430.77 A

33. 24,414.55 CM

35. 240.4 V

UNIT 37   OHM'S LAW FORMULAS

1. 0.545 A
3. 16.70 A
5. 12 Ω
7. 1.37 A
9. 0.545 A

11. 240.01 ft
13. 2,637.2 ft
15. 11.2 Ω
17. 0.6 Ω
19. 7.33 Ω

21. 0.25 Ω
23. 0.936 Ω
25. 101,323.64 CM

UNIT 38   POWER FORMULAS

1. 3600 W
3. 0.625 A
5. 1800 W

7. 114.94 V
9. 113.64 V
11. 3.33 A

13. 28.8 Ω
15. 2,332.27 W

UNIT 39   COMBINED PROBLEMS ON FORMULAS

1. 2.4 A
3. 20 Ω
5. 0.96 W

7. 31.5 ft
9. 0.548 A
11. 4.3 A

13. 691.2 W
15. 19.427 A
17. 4,800 Ω

SECTION 9   TRIGONOMETRY

UNIT 40   RIGHT TRIANGLES

1. 19.21
3. 51.96

5. 42.55
7. 59.24 ft

9. 12.17 mi

UNIT 41   TRIGONOMETRIC FUNCTIONS

1. 45.6°
3. 31.3°
5. 16.8 ft

7. 1.3 ft
9. a.   20.6°
   b.   159.4°

11. 27 ft 10 ½ in

UNIT 42   PLANE VECTORS

1. 36.06 Ω
3. 17.32 Ω
5. 22.80 V

7. 250 V
9. 127.37 V
11. 120.2 V

13. 5 A
15. 95.3 V

UNIT 43   ROTATING VECTORS

1. 8.68 A

3. 160 V

5. 86.58 A

7. a.   0 V
    b.   51.76 V
    c.   173.2 V

    d.   193.19 V
    e.   200 V

9. 100 kHz

UNIT 44   COMBINED PROBLEMS IN TRIGONOMETRY

1. 8.94

3. 287.23

5. 66.66

7. 77.9 Ω

9. 12.5 Ω

11. 207.8 V

13. 15.5 A

15. 120°